Mediterranean Ecogeography

Mediterranean Ecogeography

Harriet D. Allen

Director of Studies and Senior Lecturer, Homerton College, Cambridge
Affiliated Lecturer, Department of Geography, University of Cambridge

An imprint of **Pearson Education**

Harlow, England · London · New York · Reading, Massachusetts · San Francisco · Toronto · Don Mills, Ontario · Sydney
Tokyo · Singapore · Hong Kong · Seoul · Taipei · Cape Town · Madrid · Mexico City · Amsterdam · Munich · Paris · Milan

Pearson Education Limited
Edinburgh Gate
Harlow
Essex CM20 2JE
England

and Associated Companies throughout the world

Visit us on the World Wide Web at:
http://www.pearsoneduc.com

First published 2001

ISBN 0 582 40452 5

British Library Cataloguing-in-Publication Data
A catalogue record for this book is available from the British Library

Library of Congress Cataloging-in-Publication Data
Allen, Harriet D.
  Mediterranean ecogeography / Harriet D. Allen.
      p.   cm. — (Ecogeography series)
    Includes bibliographical references (p.   )
    ISBN 0–582–40452–5
    1. Mediterranean Region—Geography.   2. Landscape ecology—Mediterranean Region.
  3. Agricultural geography—Mediterranean Region.   4. Land use, Rural—Environmental
  aspects—Mediterranean Region.   I. Title.   II. Series.
  D973.A42   2000
  508.3182′2—dc21   00–045310

10 9 8 7 6 5 4 3 2 1
05 04 03 02 01

Typeset by 35 in 11/12pt Adobe Garamond
Produced by Pearson Education Asia Pte Ltd.
Printed in Singapore

Contents

List of plates

List of figures

List of tables

Series preface

Ecogeography marries ecology with geography. It places ecosystems in a spatial context. Its practitioners – ecogeographers – investigate the distribution and structure of ecosystems at various scales, with particular attention given to the spatial and dynamical relations among biotic and abiotic components of ecosystems. They usually focus on the 'landscape scale', defined as areas from tens of square metres to about 10 000 square kilometres, or the size Cheshire, England. However, ecogeographical studies cover all spatial scales, from a few square metres to the entire globe. Regional-scale investigations view the 'big picture', probing the distribution, structure and dynamics of ecosystems in Cheshire-sized regions to continents and the entire planetary surface. The aim of the Ecogeography Series is to provide discursive accounts of the ecogeography of the world's major ecological regions – tropical, temperate, desert, Mediterranean, polar and subpolar, mountain, riverine, lacustrine, subterranean, urban, coastal and marine. Zonal climates circumscribe some of these 'regions', but the rest are distinctive in other ways.

Each book in the series examines the nature of an ecological 'region'. Individual authors are free to bring their own slant to content, their own predilections to emphasis, but they all prosecute the ecogeographical theme. In doing so, they cover such topics as climate, topography, soils, plants and animals, communities, ecosystems, land use and environmental problems. Note that the human response to the 'regions' is an important element of the books. The Ecogeography Series is about people and environment, and is not a set of regional physical geographies.

Books in the Ecogeography Series at once provide good depth and breadth of coverage. Breadth comes from an outline of where the ecological 'regions' are found and what they are like. Depth comes from selected examples and case studies, and from carefully chosen suggestions for further reading. The books are suitable for geographical and environmental courses worldwide, and fill a gap in the market – there is, for the most part, a lack of undergraduate books with a regional focus on the people and environment theme. They are designed for students' needs and pockets, providing second-year and third-year undergraduates with weighty but wide-ranging volumes on specific ecological 'regions' at an affordable price.

Richard Huggett
June 2000

Author's preface and acknowledgements

This book is a contributory volume to a series on the *Ecogeography* of different world environments. The series attempts to 'marry' ecology with geography. In order to be successful geography has to be covered in its widest sense – the inclusion of both physical and human components of the subject – and arguably this is particularly important in the Mediterranean region. The Mediterranean has a distinctive climate, has been inhabited for thousands of years and has seen the expansion and contraction of several major civilisations. Recent decades have also witnessed massive population movements, increased urbanisation and industrialisation, and the development of tourism. Pressures on the natural resources of the region have therefore increased. Inevitably, questions have arisen about the relative impacts of recent changes compared with those which occurred through prehistory and the historic period ending in about 1945. As in other parts of the world this has led to a need and demand for environmental protection measures to prevent further resource degradation. But we need to ensure that any such measures are informed by a thorough understanding of the ways in which different ecosystems in the Mediterranean region operate. It is becoming apparent that in some environments, especially those in upland rural areas, plant and animal communities have developed alongside thousands of years of land-use practices. When such practices, such as grazing or non-intensive agriculture, are abandoned landscapes change and not always for the better in terms of the survival of plant and animal species. However, it is wrong to assume that land-use practices and resource management in one part of the Mediterranean region can be directly applied to another region. The Mediterranean is topographically and culturally diverse and this needs to be taken into account in any volume that examines the region's ecogeography. To date, many books on the Mediterranean region have focused on either its society or its history or its ecology. It is hoped that this volume achieves the aim of integrating its ecology and geography.

There are in fact five regions of the world with mediterranean-type climates – the Mediterranean Basin itself, California, central Chile, the very southern tip of South Africa and south and western Australia. Arguably this book could therefore cover the ecogeography of all regions with mediterranean-type ecosystems and, indeed, much research has been undertaken with respect to the comparability or distinctiveness of such ecosystems in the five regions. However, this volume focuses on the Mediterranean Basin. While some readers might regard this as an oversight, or even a limitation, it allows the subject matter to be treated in greater depth than might have been possible (given constraints on book length) had all five regions been covered, probably in a more superficial manner. There are, I hope, sufficient references in the text to the other regions which will allow readers to follow up any interest they develop in mediterranean-type ecosystems.

I came to be interested in the Mediterranean when at school. I was lucky enough to have a French exchange pen-friend who lived just outside Aix-en-Provence. Staying with her introduced me to far more than the French language – a different way of life, different food, different history, different climate, different landscape etc. This early interest was then fostered by other holiday visits to different Mediterranean countries. My academic interest then built on this through the opportunity to pursue a PhD at Cambridge with Dick Grove as my supervisor. For me this married the Mediterranean with an interest in Quaternary science. Through several visits to Greece in the company of other geographers, botanists and archaeologists I came to understand so much more about the Quaternary history and recent past, including biogeographical and ecological aspects, of the country. Driving out to Greece also allowed comparisons with other Mediterranean countries through which we passed. More recently I have been working in the Algarve, Portugal. Going on field trips with thirty geography undergraduates, most of whom are not in the least interested in biogeography, has also made me think about ways in which to make the subject more interesting (I hope). I am therefore very grateful to all those people in Cambridge, Greece, Portugal and elsewhere who have fed my curiosity. In particular I owe a big debt of gratitude to Dick Grove for all his patient support. Dick also kindly lent me drafts of his forthcoming book (with Oliver Rackham): *The Nature of Mediterranean Europe: an historical ecology*.

I am indebted to Richard Huggett of Manchester University and Matthew Smith, Senior Acquistitions Editor, at Pearson Education Limited for the chance to write this book. Both of them have been very supportive, as has Anita Atkinson, Senior Editor, all of whom have answered many of my queries and made suggestions which I am sure have improved the book's content. I also wish to thank those who reviewed parts of the manuscript: Steve Trudgill, Dick Chorley, Tjeerd van Andel, Michael Reiss, Dick Grove and Richard Huggett. Their comments and criticisms have been immensely helpful.

A multidisciplinary book such as this one obviously requires a large amount of armchair (or library) research and I have been very fortunate to have had access to Cambridge University Libraries. In particular Jane Robinson and Colin MacLennan of the Geography Department Library and Richard Savage of the Plant Sciences Library have sought out books, journals and inter-library loans for me. Jane Robinson also read the entire manuscript to check for consistency, though any such errors which remain are mine. Information about the Indigenous Propagation Project (referred to in Chapter 9) came from Sarah Paul and Abigail Entwistle at Fauna and Flora International. My thanks go to them for this. I could not have found the time to write this book were it not for the support of Homerton College, Cambridge. The college provided some study leave and my colleagues and students put up with my sporadic availability. Also at Homerton, I was assisted in the preparation of the diagrams by Janet Rogers. I am also very grateful to those publishers who gave permission for copyright material to be reproduced in the book. Acknowledgements for these are given in 'Publisher's acknowledgements' with source details by the relevant figures and supported in the bibliography. Finally the writing of this book has only been possible through the tolerance, love and support of Ian, Frances and Isabel.

Harriet D. Allen
June 2000

Publisher's acknowledgements

We are grateful to the following for permission to reproduce copyright material:

Faber & Faber Limited/authors' agents Curtis Brown Ltd, London, reproduced with permission, on behalf of the Estate of Lawrence Durrell for an extract from *THE GREEK ISLANDS*. Copyright Lawrence Durrell 1978.

Figure 1.1a, Figure 6.1a, Figure 6.3a and Figure 6.3b from di Castri, F., Goodall, D.W. and Specht, R. (eds) (1981) *Ecosystems of the World II Mediterranean-type Shrublands*, copyright 1981, and Table 4.2 from Kosmos, C. *et al.* (1997) 'The effect of land use on runoff and soil erosion rates under Mediterranean conditions' *Catena* 29, pp. 45–59, copyright 1997, reprinted with permission from Elsevier Science; Figure 2.1c from UNESCO-FAO (1963) *Bioclimatic map of the Mediterranean Zone*, reprinted with permission from UNESCO, and Kraus, H. and Alkhalaf, A. (1995) 'Characteristic surface energy budgets for different climate types' *International Journal of Climatology* 15, reprinted with permission of John Wiley & Sons Limited; Figure 2.5, Figure 2.6a, Figure 2.6b, Figure 2.7a, Figure 2.7b, Figure 4.3 and Figure 8.7 from Mairota, P., Thornes, J.B. and Geeson, N. (eds) (1998) *Atlas of Mediterranean Environments in Europe: The Desertification Context*, copyright 1998 © John Wiley & Sons Limited, and flow chart (in Chapter 7) from Blumler, M.A. (1993) 'Successional pattern and landscape sensitivity in the Mediterranean Near East' from Thomas, D.S.G. and Allison, R.J. (eds) *Landscape Sensitivity*, copyright 1993 © John Wiley & Sons Limited, reproduced with permission; Figure 2.7a from Palutikof, J.P., Guo, X., Wigley, T.M.L. and Gregory, J.M. (1992) 'Regional changes in the climate in the Mediterranean Basin due to global greenhouse gas warming' *UNEP Mediterranean Action Plan Technical Report Series* 66, 172, reprinted with permission from the United Nations Environment Programme; Figure 3.3a reprinted from *Mediterranean Quaternary River Environments* Lewin, J., M.G. Macklin and J. Woodward (eds). 90 5410 191 1, 1995, 25 cm, 300 pp., EUR 82.50/US$97.00/GBP58.00, and Figure 8.2 reprinted from *Man's Role in the Shaping of the Eastern Mediterranean Landscape – Proceedings of the symposium on the impact of ancient man on the landscape of the Eastern Mediterranean region and the Near East*, Groningen, 6–9 March 1989, Bottema, S., G. Entjes-Nieborg and W. van Zeist (eds). 90 6191 138 9, 1990, 25 cm, 352 pp., EUR 104.50/US$123.00/GDP 74.00, by permission of A.A. Balkema, P.O. Box 1675, Rotterdam, Netherlands (fax: +31.10.4135947; e-mail: sales@balkema.nl); Figure 3.3b from Dedkhov, A.P. and Moszherin, V.I. (1992) 'Erosion and sediment yield in mountain regions of the world' from *Erosion, Debris Flow and Environment in Mountain Regions* (Proceedings of the Chengdu Symposium, July 1992), IAHS Publ. No. 209, reprinted by permission of IAHS Press; Figure 3.5a and Figure 3.5b from van Andel, T.H. (1989) 'Late Quaternary sea-level changes and archaeology' *Antiquity* Vol. 63 and Figure 8.5b from van Andel, T.H. and Runnels, C.N. (1995) 'The earliest farmers in Europe' *Antiquity* Vol. 69, reprinted by permission of the authors and Antiquity Publications Ltd; Figure 4.1, Figure 4.2 and Table 4.1 from Bridges, E.M. (1978) *World Soils*, Second Edition, and Table A1.1 from Stace, C.A. (1991) *New Flora of the British Isles* and Tutin, T.G. *et al.* (1964–1980) *Flora Europaea*, 5 Volumes, reprinted with permission of Cambridge University Press; Figure 5.1 from Goudie, A.S., Atkinson, B.W., Gregory, K.J., Simmons, I.G., Stoddart, D.R. and Sugden, D. (eds) (1994) *The Encyclopedic Dictionary of Physical Geography*, Second Edition, and Figure 8.1 and Table A3.1 from Roberts, N. (1998) *The Holocene*, Second Edition, reprinted by permission of Blackwell Publishers Ltd; Figure 5.4a and Figure 5.4b from Esau, K. (1965) *Plant Anatomy*, Second Edition, copyright © 1965 John Wiley & Sons, Inc, reprinted by permission of John Wiley & Sons, Inc; Figure 6.1b from Mazzoleni, S., Lo Porto, A. and Blasi, C. (1992) 'Multivariate analysis of climatic patterns of the Mediterranean basin' *Vegitato* 98, pp. 1–12, Figure 6, reprinted with kind permission from Kluwer Academic Publishers; Figure 6.2 from King, R., Proudfoot, L. and Smith, B. (eds) (1997) *The Mediterranean: Environment and Society* and Figure 6.4 from Biger, G. and Liphschitz, N. (1991) 'The recent distribution of *Pinus brutia*: a reassessment based on dendroarchaeological and dendrohistorical evidence from Israel' from *The Holocene* 1, reprinted by permission of Arnold Publishers; Figure 7.2 from Noss, R.F. (1990) 'Indicators for monitoring biodiversity: a hierarchical approach' *Conservation Biology* 4, reprinted by permission of Blackwell Science, Inc; Plate 7.1 © M. Biancarelli/Woodfall Wild Images; Plate 7.2 © Alexis Maryon; Figure 7.3 from Médail, F. and Quézel, P. (1997) 'Hot-spots analysis for conservation of biodiversity in the Mediterranean Basin' *Annals of the Missouri Botanical Garden* Vol. 84, reprinted by permission of Frédéric Médail; Figure 7.4 from Mooney, H.A. and Dunn, E.L. (1970) 'Convergent evolution of Mediterranean-climate evergreen scleophyllous shrubs' *Evolution* 24, reprinted by permission of Society for the Study of Evolution; Plate 8.2 © Royal Geographical Society, London; Plate 8.3 © ESA (2000) distributed by Eurimage/HME Partnership, Romford; Figure 8.3 from Hillman, G., Colledge, S.M. and Harris, D.R. (1989) from Harris, D. and Hillman, G. (eds) *Foraging and Farming* (pub Unwin Hyman) reprinted with permission of Routledge; Figure 8.6 from van Andel, T.H., Zangger, E. and Demitrack, A. (1990) 'Land use and soil erosion in prehistoric and historical Greece' *Journal of Field Archaeology* 17, reprinted by permission of the Journal of Field Archaeology; Table 9.1 from IUCN Species Survival Commission (1994) *IUCN Red List Categories*, reprinted by permission of IUCN – The World

Conservation Unit; Plate 9.2 © David Woodfall/Woodfall Wild Images; Plate 9.3 © Pascal Pernot/Still Pictures; Table A2.1 from Harland, W.B. (1990) *A Geological Time Scale*, reprinted with permission of the author and Cambridge University Press.

Chapter 1

The Mediterranean: an introduction

To someone flying into Faro airport on a sunny winter or spring afternoon the fields of the Algarve below appear swathed in yellow. Down on the ground the culprit turns out to be a small, not especially pretty, yellow flowered plant – the Bermuda buttercup (Plate 1.1). It is the culprit because this invasive weed is toxic to cattle; it contains high levels of oxalic acid. It has also spread widely, not just through Portugal but elsewhere in the Mediterranean. *Oxalis pes-caprae*, to give it its proper name, came from southern Africa as a garden ornamental but quickly became naturalised. It was first introduced to Sicily in 1796 and it arrived in Portugal in 1825. It was probably introduced several times in different places because it is so easy to propagate. This fact also makes it very difficult to eradicate.

Oxalis pes-caprae is just one of many hundreds of exotic plant species found in the Mediterranean. Plants which we think of as typically 'Mediterranean' often turn out to be exotic – think of Mediterranean cuisine with its reliance on tomatoes and capsicums (peppers) from the Americas and aubergines from

Plate 1.1 *Oxalis pes-caprae* (Bermuda buttercup), an exotic species introduced to the Mediterranean region from southern Africa in the eighteenth century.

tropical Asia. Not all introduced species were brought deliberately as food, for horticulture or as crops. Many were accidental arrivals such as those of the *Flora Juvenalis* of Montpellier in France. The building of the Lez canal in 1686 meant that wool stores could be established at nearby Port Juvenal. Wool was imported from many exotic locations with seeds trapped in the fibres. As the wool was washed the seeds were freed and some germinated successfully. With repeated introductions firm colonies became established. By 1813, 13 exotic species were flourishing; there were 386 species in 1853 and as many as 458 in 1859. But in 1880 storing and washing of wool ceased and gradually the number of exotics declined to only six in 1950. Environmental conditions for the germination of seeds from plants growing in nearby meadows were quite different from the warm waters of the wool processing. No new seeds were brought in and so the colonies were no longer able to survive. As a result most of these accidental introductions did not become naturalised or invasive (Groves, 1991).

Some exotic flora are easier to recognise than others as trees or garden escapees familiar from gardens around the world. The streets of many Mediterranean towns are lined with azure-coloured jacaranda trees (for example *Jacaranda mimosifolia*) or the aptly-named bead-trees or Persian lilacs (*Melia azedarach*) with hard fruits once used for rosary beads hanging on bare stems in winter. Plane trees (*Platanus* spp.) or sweet-smelling citrus trees often provide the shade for sitting outside tavernas, ristorantes or cafés, and palm trees (*Phoenix* spp.) line beach fronts. If you start to ask where these come from then you start to delve into the history of Mediterranean people and their trading routes. Some of these trees go back to prehistory – citrus trees were originally native to Asia and have been cultivated and hybridised for so long that their origins are lost. It is assumed that the citron (*Citrus medica*) was introduced by Alexander the Great (356–323 BC) to Greece and the Near East from India. Theophrastus, writing in Greece in the fourth century BC, certainly described their cultivation. However, an earlier date is possible as a citrus seed has been found at archaeological excavations in Cyprus dated as early as 1200 BC. Some plants have cultural associations with particular groups of people. The Moors swept through the Iberian Peninsula from the eighth century AD and are believed to have taken with them many new crops: rice, lemons, apricots, cotton, bananas, watermelons and aubergines are all listed in the Calendar of Cordoba, of about AD 961. Trees from the New World and the Pacific came after the great fifteenth- and sixteenth-century AD voyages of exploration, such as the Norfolk Island pines (*Araucaria* spp.) from the Pacific. Other plants came from the east with traders travelling into China. Today, looking at these trees should be like looking at an ancient ruin or an architectural marvel – it should bring to mind the people who lived in the landscape and planted the vegetation, just as buildings bring to mind people and their history.

Unravelling and understanding a landscape is difficult but it is helped by the clues that the vegetation has to offer. While thousands of species have been introduced, the 'native' flora itself is huge. The Mediterranean Basin has one of the richest in the world and many species are endemic, found only in restricted localities. Knowing the ecological preferences of these plants helps to make sense of plants and places – why do plants live where they do? On the Venetian walls of Khania in Crete *Petromaroula pinnata* throws out spikes of blue flowers in May and June. This species is only found on Crete, and is common on cliffs, except in the driest areas (Rackham and Moody, 1996). Why does it grow only in these places? Is the answer to do with microclimate, or soil type, or escaping from being browsed or grazed? Sheep and goats seem to be ubiquitous in the rural Mediterranean – their bells ring out across the hillsides – and there is a long history of blaming the avaricious goat for eating too many plants too widely, causing long-term degradation and erosion of hillslopes. Yet goats and sheep have lived alongside people for thousands of years since their domestication in the Near East. Surely by now landscapes and plants have had time to adjust to their presence, and the animals should have become a part of the way landscapes work, in other words a part of some Mediterranean ecosystems?

This book attempts to answer the questions raised in these introductory paragraphs. This is the realm of ecogeography – the marriage between geography and ecology. This is a study of the relationships between plants and animals and their environment – the influence of climate, geomorphological processes, soils and especially people. In one sense this is akin to the geographer's meaning of biogeography – the study of the biosphere and of human effects on plants and animals. It is, however, different from a biologist's understanding of biogeography – the study of past and present distributions of plants and animals at various scales and how these distributions came about.

The Mediterranean Basin has a long history of settlement and therefore human impact and this has increased rapidly in recent decades. Through the twentieth century, and particularly the second half, the economy of the Mediterranean region changed dramatically. The processes of industrialisation and urbanisation, and a massive growth in tourism have put pressures on the natural and semi-natural landscapes. Despite the problems these have caused – draining of wetlands, polluted water courses, over-abstraction of groundwater, for instance – it would be wrong to view these economic changes as uniform across the region. Not all parts of the Basin have even been equally affected and in the same ways – Mediterranean geography is one of uneven development, just as its history is varied across the Basin. These factors need to be considered in any explanation of Mediterranean ecogeography.

1.1 What do we mean by 'Mediterranean'?

How do we define 'Mediterranean'? What criteria can we use to identify the Mediterranean region? There is no easy answer to this. For some people it is the nature of its climate that makes the Mediterranean distinctive – a region of warm wet winters and hot dry summers. Or it could be a region with a

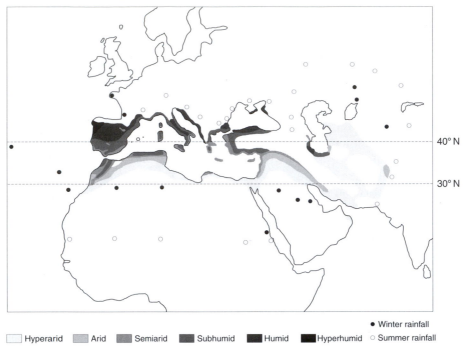

| | Hyperarid | | Arid | | Semiarid | | Subhumid | | Humid | | Hyperhumid | ○ Summer rainfall |

● Winter rainfall

Figure 1.1a   Map of the Mediterranean climate region. Stations surrounding the Mediterranean-climate zone with predominating winter or summer precipitation are also shown. After di Castri (1981).

distinctive flora, perhaps the distribution of the olive, or of evergreen sclero-phyllous shrubland. For others it is a geographical region encompassing all those countries with a coastline on the Mediterranean Sea. For others still, the word has a cultural connotation. This might be in relation to its food, or its histories, with a degree of shared civilisations – Greek, Phoenician, Roman, Byzantine, Moor or Ottoman. Each of these criteria has its problems.

Although Mediterranean climates are characterised by hot dry summers and warm wet winters there is also marked variability within, as well as between, seasons. In summer strong dry winds may blow, but not all the time, while in winter there may be strong cold winds. Local climates are much modified by relief (orographic and rainshadow effects), aspect, altitude and distance from the sea. In general, frosts are rare but do occur, especially in mountainous areas. If we collect data to illustrate these characteristics then it is clear that while such a climate is found around the Mediterranean Sea it also extends to the east as far as western Pakistan (Figure 1.1a). Four other parts of the world also have a similar climate – the Cape Region of South Africa, western and south Australia, California and central Chile (see Box 1.1). These regions typically have 90% or more of their annual precipitation falling in the six cool

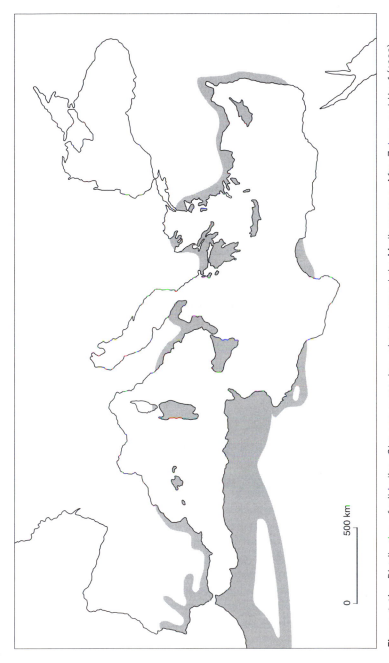

Figure 1.1b Distribution of wild olive, *Olea europaea* subsp. *oleaster*, around the Mediterranean. After Zohary and Hopf (1993).

(continued)

Table 1.1 Plant species diversity, topographic and climatic diversity, and regional disturbance threats in mediterranean-climate regions of the world. Order of regional disturbance threats roughly follows their significance, beginning with the most critical factors

Region	Area (10^6 km^2)	Native plant species	Topographic heterogeneity	Climatic heterogeneity	Regional disturbance threats
Mediterranean Basin	2.30	25 000	High	Very high	Deforestation, overgrazing, agriculture, urbanisation
California	0.32	4 300	High	Very high	Urbanisation, agriculture
Central Chile	0.14	2 400	Very high	Very high	Deforestation, overgrazing, agriculture
Cape region, South Africa	0.09	8 550	Moderate	High	Invasive alien plants, agriculture, urbanisation
Southwestern Australia	0.31	8 000	Low	Moderate	Agriculture, deforestation, introduced pathogens

Source: Rundel (1998).

One of the most widely recognised symbols of the Mediterranean world is the olive (Plate 1.2). It grows only in typical Mediterranean environments, and its cultivation, and the trade in its fruit and oil, has accompanied the spread of people around the Basin from Bronze Age times. The cultivated olive (*Olea europaea*) has a close affinity to wild olives (*Olea europaea* subsp. *oleaster*) and even today the wild olive thrives as a constituent of Mediterranean *maquis* and *garrigue* vegetation communities (Zohary and Hopf, 1993). Its distribution is shown for comparison with a Mediterranean climate in Figure 1.1b. But of course the olive is not the only characteristically Mediterranean plant. Others that we might recognise are the holm oak (*Quercus ilex*), the fig (*Ficus carica*), the Aleppo pine (*Pinus halepensis*) and the mastic or lentisc tree (*Pistacia lentiscus*).

The holm oak and mastic tree, together with wild olive, are generally recognised as maquis plants, and maquis communities are themselves seen as typically Mediterranean. The word maquis has come to describe an impenetrable cover of dense vegetation, composed of small trees and bushes of medium height with evergreen leaves. Sometimes it is distinguished from

Plate 1.2 Cultivated olive tree growing on the edge of a terrace with further terraces beyond.

similar garrigue communities which will grow on calcareous rocks, whereas maquis is confined to siliceous substrates, but not everyone makes this distinction and the word maquis is used for both. Indeed it has come to mean more than just a vegetation community but a whole landscape incorporating people and their responses to it (Perelman, 1981). The word has its origins in the Corsican dialect where *macchia* means 'that which is mottled' and in the Corsican tradition people 'took to the maquis' to escape the law. During the Second World War the resistance movement in France was known as the Maquis and its fighters as the *Maquisards*. The impenetrability of the maquis makes it a labyrinth for walkers, who can barely see their way, as well as a refuge for those in hiding.

Geopolitically the Mediterranean comprises those countries bordering the Mediterranean Sea but this is a fairly restrictive definition. It excludes Portugal and parts of Jordan – mediterranean in climate and culture, but not in coastline – but includes countries which have vast tracts of land beyond the immediate influence of the Sea, notably Algeria, Libya and Egypt, which extend far south into the Sahara desert, and France with its North Atlantic coastline and borders with northern European nations such as Belgium and Germany. The break-up of the former Yugoslavia has also produced states with variable links to the Mediterranean Sea. Possession of a common Mediterranean coastline formed the basis of the 'Blue Plan', a discussion on futures for the Mediterranean Basin (Grenon and Batisse, 1989). For practical purposes this adopted the coastal administrative divisions of each country bordering the Sea in order to plan a common approach to environmental issues associated with water shortages, tourism and urbanisation. It was argued that the administrative districts are largely focused around the Sea itself but given the way in which these regions are delimited, quite small but densely populated administrative districts, in for example France and Italy, are contrasted with featureless, coastal wastes of parts of Egypt and Libya (King, 1997a).

For the purposes of this book no single definition of the word Mediterranean is applied. However, given its emphasis on ecogeography the defining criteria are based on combinations of climate and ecology. This reflects the research and literature in these fields. It means that Portugal is included but Libya and Egypt are largely excluded, though there are some references to the Nile Valley.

There are very few, if any, areas of the Mediterranean which do not bear an imprint of human activity. Therefore it would be wrong to regard the landscape or its component ecosystems as natural. Some upland areas, coastal dunes and wetlands may be the last refuges of natural, spontaneously occurring and reproducing plants and animals (Naveh, 1995) but it makes sense to regard many Mediterranean ecosystems today as semi-natural. Although this is a rather vague term in itself, being everything in between natural and artificial, it is the semi-natural landscapes that are the focus of this book. These are generally managed landscapes or ecosystems, as in the *dehesa* woodlands of Spain and Portugal or the olive groves widespread around the Mediterranean. The role of agriculture is therefore an important consideration in their ecogeography. However, what this book does not consider in depth is modern, intensive agriculture unless it impinges on other ecosystems or environmental concerns.

1.2 Historical overview

The Mediterranean Sea is clearly the focus of the Mediterranean region and around it several great civilisations have emerged and spread, often beyond its shores. The Sea is virtually tideless and the predictable regularity with which winds blow during certain seasons allowed trade to flourish and cultures to spread. The Sea is also the meeting place of Europe, Africa and Asia and so inevitably people and their ideas from these three continents met and diffused around the coast. The climatic influence of the Sea and the geological similarity of the continents around it lend a sense of unity to the region, but look closely and its diversity becomes clearer. The Sea separates largely Christian Europe from the Islamic world of North Africa and the Middle East and so reflects the history of the region. In turn, this has affected the ways in which the land has been used, both today and in the past, and consequently the natural history of the Mediterranean.

To understand the ecogeography of the Mediterranean an overview of the region's history and modern economy is given (see Table 1.2 for demographic data on Mediterranean countries). By necessity this is brief and therefore superficial. It does not seek to interpret history, explaining why events happened or their consequences. Other far better books exist for those purposes (see the suggested reading at the end of this chapter). Neither is it the intention of this overview to suggest that the impact on the landscape of any particular group

Table 1.2 Demographic, socio-economic and land-use data for the countries of the Mediterranean

Country	Area ×1000 sq km [a]	Area of MTE (%) [b]	Population (1996) [a]	Population growth rate, 1980/96 (%) [a]	GNP per capita $ (1996) [a]	EU member [a]	Labour force in agriculture (%) [a]	Arable land	Permanent crops	Permanent as % of total land area	Forest and woodland	Other land	Irrigated land as % arable and cultivated land
Albania	28 748	72	3 401 000	1.5	820		nd	21	5	15	38	22	60
Algeria	2 381 741	9	28 784 000	2.7	1 520		26	3	0	13	2	82	5
Croatia	88 117	nd	4 501 000	nd	nd		16	nd	nd	nd	nd	nd	nd
Cyprus	9 251	100	756 000	1.3	nd		14	12	5	1	13	69	23
Egypt	1 001 449	0	63 271 000	2.3	1 080		nd	2	0	0	0	97	100
France	551 500	12	58 333 000	0.5	26 270	1957	5	33	2	20	27	18	6
Greece	131 990	92	10 490 000	0.5	11 460	1981	23	22	8	41	20	9	31
Gibraltar		100	28 000	nd	nd		nd	nd	nd	nd	nd	nd	nd
Israel	21 056	<5	5 664 000	2.4	15 870		4	17	4	7	6	66	41
Italy	301 268	58	57 226 000	0.1	19 880	1957	9	31	10	17	23	20	26
Jordan	97 740	nd	5 581 000	4.0	1 650		15	4	0	9	1	86	16
Lebanon	10 400	>95	3 084 000	0.9	2 970		7	21	9	1	8	61	28
Libya	1 759 540	<1	5 593 000	3.8	nd		11	1	0	8	0	91	11
Malta	316	100	369 000	nd	nd		3	37	3	0	0	59	8
Morocco	446 550	33	27 021 000	2.1	1 290		45	20	1	47	20	12	14
Portugal	91 820	>95	9 808 000	0	10 160	1986	18	26	9	9	32	24	20
Spain	505 992	66	39 674 000	0.3	14 350	1986	12	31	10	21	32	7	17
Syria	185 180	32	14 574 000	3.2	1 160		33	26	4	42	4	23	12
Tunisia	163 610	35	9 156 000	2.2	1 930		28	19	13	28	4	36	5
Turkey	774 815	60	61 797 000	2.1	2 830		53	32	4	11	26	27	9
Fed. Rep. Yugoslavia	102 173	nd	10 294 000	nd	nd		nd	nd	nd	nd	nd	nd	nd

nd, no data.

Sources:

[a] *The World Guide 1999–2000*, New Internationalist Publications, Oxford.

[b] Area of Mediterranean-type ecosystems from Hobbs *et al.* (1995), but various calculations are available dependent on climatic criteria.

[c] Dunford (1997); data are for 1991 (except Jordan and Portugal, 1986–88).

of people, such as the Romans or Moors, was equal in all regions. Different areas and sectors of people and their economies were influenced in different ways. We also need to remember that within any civilisation there were schisms, for example between the Latin and Orthodox Christian churches. Therefore to talk, for example, of a 'Christian' world is itself a generalisation.

1.2.1 North Africa and the Near East

The countries of North Africa have a distinctively un-African character as it is the Sahara which forms the real boundary between tropical Africa and the land masses to the North. In this sense North Africa is a Mediterranean region, but one belonging to the Arab world, though earlier colonisers in modern Tunisia were the Phoenicians and Romans. The Phoenicians founded Carthage near present-day Tunis and the Romans then governed the region, with some interruptions, until the arrival of the Arabs. From the seventh century AD Arab and Berber invasions brought hundreds of thousands of people to the Maghreb – Gezira-el-Maghreb, or the Island of the West – the countries of Morocco, Algeria and Tunisia. They brought with them the agricultural economies of the Near East – wheat and olives became the predominant crops of the valleys and coastal plains. In the mountains livestock rearing has been important. Cultivation of figs and olives is widespread. From the Maghreb the Arabs, or Moors, crossed the Strait of Gibraltar in AD 711 to colonise the Iberian Peninsula. They then retreated back to the Maghreb with the Christian reconquest of the Peninsula from AD 1140 onwards.

Arab and Berber dynasties ruled the Maghreb until parts of the region were conquered by the Turks in the sixteenth century. In turn these were replaced by European colonists from the nineteenth century – the French and Spanish. In 1830 Algeria became a part of France ruled directly from Paris, while Tunisia and Morocco were governed indirectly by the French Ministry of Foreign Affairs under the nominal sovereignty of local rulers. From 1912 the Spanish and French divided Morocco. The Spanish protectorate flanked the Strait of Gibraltar, while the rest of the country was French. Not until 1956 did the country gain independence, though even today there are Spanish enclaves at Ceuta and Melilla on the Mediterranean coast. Algeria became an independent republic in 1962 and Tunisia was granted independence in 1956.

The contribution of agriculture to the economies of the Maghreb countries has changed (Pratt and Funnell, 1997; Dunford, 1997). In the immediate post-colonial period about half of all employed people worked in the agricultural sector (Algeria 57%, Tunisia 49% and Morocco 61% in 1965). Most were sedentary peasants producing cereals, some tree crops in mountain areas and market gardening in irrigated regions near to cities. Pastoralism, which had been important previously, was in decline. By the early 1990s employment in agriculture had declined considerably (Algeria 18%, Tunisia 26% and Morocco 46% in 1992). A shift from a rural to an urban population has accompanied this decline. In Algeria some 52% of people live in cities and in

Morocco and Tunisia 46% and 56% respectively. Yet the countryside is more densely populated than ever because of population explosions in all of the region, due to a marked reduction in death rates while birth rates have remained high. The annual average population growth for 1961–1992 was between 2% and 3% in each of the Maghreb countries, resulting in a practical trebling of population since 1950 (Laouina, 1998). Agricultural modernisation has occurred and production increased, but production per head has declined. The countries are increasingly dependent on food imports and/or further expansion of output. As large areas are uncultivable because of aridity (for example 82% of Algeria is uncultivated) this means further modernisation. The area of irrigated land has expanded and the range of crops grown has broadened. Vegetables and fruit, especially citrus fruits, olives and tobacco, have expanded. Some is produced for the home market, the rest is destined for export, particularly to the European Union market. For example, tomato and strawberry growing in the Souss Valley in Morocco increased massively through the 1980s and 1990s. However, many crops have low yields. By comparison with France, where a hectare of wheat yields 67 quintals, the figures for Morocco, Tunisia and Algeria are 19, 17 and 11 quintals respectively. Reasons for low yields include small field sizes owned by farmers producing on small farms, limited rainfall (most cereal production is rain-fed rather than irrigated), cultivation on steep slopes and accompanying soil erosion, and a high percentage of fallow. There is also an under-emphasis on appropriate seed selection and use of modern agronomic practices (Pratt and Funnell, 1997).

Government investment to increase agricultural output has taken place. During the colonial period this was targeted at settlers. After independence, policies aimed to modernise agriculture by trying to stimulate supply without increasing prices. However, food self-sufficiency has continued to decline and agricultural development needs to be set within a wider context of socio-political movements in these countries. In particular, there is a very marked contrast between urban affluence and rural poverty and consequently increasing social unrest and political instability (Joffé, 1993). In Algeria, for example, this unrest has exacerbated the decline in agricultural output and incomes. In addition trade relationships between the Maghreb countries and the European Union have altered the potential for export. Since Portugal, Spain and Greece joined, the EU has become self-sufficient in several products to which formerly Morocco, Algeria and Tunisia contributed. The EU is supposedly 100% self-sufficient in vegetables and olive oil and 76% in citrus fruit. The Maghreb countries therefore face stiff competition for their produce from lowered EU prices.

Agriculture is, of course, not the only economic activity in these countries and neither is economic activity restricted to the Mediterranean zone (Laouina, 1998). In Algeria there are vast hydrocarbon reserves in the Sahara which are piped to the coast and there form the centre of a petrochemicals industry. To some extent this has been at the expense of agriculture, for the Mediterranean coastal belt is the only area suited to cultivation. In Morocco, the non-agricultural sector is dominated by phosphate production. This country is the world's leading exporter of phosphates, but their importance to the economy has fallen as prices have declined. In Tunisia the industrial sector does not yet compete with agriculture for economic importance. Instead of developing industry there has been a focus on expanding tourism and this sector of the economy is growing rapidly. At present an insufficient number of jobs has been created for the growing, youthful population and as there are limited opportunities for rural migrants to find work in cities many have left for mainland Europe. In the early to mid-1980s there were some 4000 Moroccans with residence permits living in Spain, and 2600 in Italy. By 1993 these figures had increased to 61 300 in Spain and 97 600 in Italy (King, 1997b). In both countries there were more Moroccans than immigrants from any other country.

The pressures to produce food have somewhat inevitably led to environmental degradation. In particular there has been a major decline in forest cover. In Morocco present rates of forest clearance are some 20 000 to 25 000 ha annually, much of it for firewood (Laouina, 1998). Estimates of annual firewood cutting are put at three times the rate of renewal. Some forest, about 4500 ha, is cleared directly for extension of cultivation. Much of this firewood cutting and forest clearance is illicit and the result of poorly conceived and applied legislation. In consequence forest management and profitability are suffering. Economically the most important species are cedar, fir and pine – all of which are threatened in the short term. There is some reforestation but rates are very low, about 7.8%, though higher figures are achieved in more humid regions. Forest decline is also believed to be the result of overgrazing and the more frequent occurrence of fires. Secondary regrowth on cleared, grazed and burned land is producing a new and different landscape (Barbero *et al.*, 1990).

The area of the Near East with a mediterranean-type climate is restricted to the coastal zone of Anatolia in Turkey, the eastern Mediterranean coastal plain and the adjacent mountain ranges, and the island of Cyprus. The region is generally recognised as part of the Fertile Crescent – one of the ancient locations for domestication of plants and animals – wedged between the Mediterranean Sea and the Syrian and Arabian deserts. It has been densely populated for thousands of years, with most people living on the coast with its mediterranean-type climate; about 80% of Syria's population live in the 20% of the country which is Mediterranean. Throughout history the region has been a meeting point for the peoples of Africa and Asia.

The Anatolian coastline was part of the area of original Greek city-state, or *polis*, development between the eighth and sixth centuries BC. In the fourth century BC Alexander the Great's empire included the Near East. Following his death the empire broke up and the Middle East was divided between different Hellenistic empires – the Ptolomies ruled Egypt and parts of the Levant, and the Seleucids ruled in Turkey, Syria and Afghanistan. Although wars between these kingdoms continued they survived until their incorporation into the Roman empire between the late second century BC and the first century AD (Proudfoot, 1997a). The region remained Roman until after the empire broke up into eastern and western sectors. Constantinople, founded in AD 330 on the site of the old Greek town of Byzantium, became the centre of the Eastern Empire. The Byzantine empire administration of the region was much the same as it had been under the Romans until the Arab invasions. By AD 650 Egypt and the Levant as far as Antioch had fallen to the Arabs. In AD 1453 the Ottoman empire had replaced the Byzantine empire in Constantinople.

The Ottomans ruled with a tendency towards devolution and local administrative autonomy and their empire's economy was largely agrarian. Wars with Russian and European armies ensured changing territorial boundaries and shifting populations (Proudfoot, 1997b). The loss of the Balkans in the 1820s resulted in accelerating migrations of Muslim refugees to Anatolia and Syria. By 1913 an estimated 5 to 7 million refugees had arrived, adding to pressure on resources. The Ottomans lost control over Cyprus in 1878. In Turkey itself the revolution of 1908 overthrew the Ottomans and brought in a reformist and later (1923) republican government. At the end of the First World War, the remaining Turkish possessions in the Levant were divided between the French – Lebanon and Syria, both gaining independence in 1946 – and the British – Jordan, which gained independence in 1946, and Palestine, with the subsequent creation of the Israeli state in 1948.

1.2.2 The European Mediterranean

Economically France and Italy are at the centre of development in the Mediterranean, surrounded by countries of progressively lower levels of development, as shown in Table 1.2. This partly reflects their role as established members of the European Union, which allowed them to develop intra-community trade links, but of course both countries are not entirely Mediterranean in character – much of their economic activity is situated to the north and is integrated with the economic development of northern Europe. Greece joined the EU in 1981 and Spain and Portugal in 1986 and have gained from the Union's policies to promote regional development. This has brought about investment in new infrastructure and in industrial and agricultural production. The other European Mediterranean countries are in differing states of applying for membership.

The contemporary economic development of the European Mediterranean is derived from its historical roots as well as its recent history. Successive civilisations have left their marks on modern nations through cultural, social, political and legal frameworks and in the archaeological legacy of many countries (Plate 1.3). In the Aegean, Greek city-states developed from the eighth century BC and by 600 BC most of these *poleis* had an urban form with a market, or *agora*, a place of assembly and a seat of justice or government (Proudfoot, 1997a). Through colonisation the Greeks took their culture to the coastal regions of Italy, southern France, the borders of the Black Sea and parts of North Africa. Here they came into conflict with Carthaginian cities of earlier, Phoenician, origin. The Carthaginians had established colonies along the modern Tunisian coast and around the Strait of Gibraltar. Trade and war between the different city-states and their neighbours spread new technologies

Plate 1.3   The ruins of Delphi on the lower slopes of Mount Parnassus, Greece. Considered by Ancient Greeks to be the centre of the world, the site was sacred to the god Apollo and visitors came from far to consult its oracle. It was an important site from the sixth to early second centuries BC. It was designated a World Heritage site in 1987.

and ideas, including agricultural practices, across the Mediterranean. This continued with the rise of the Romans and their territorial expansion through the region.

By the late sixth century BC the Romans had established their supremacy around Rome itself and then began to expand. By 218 BC the Romans controlled peninsular Italy and its contiguous regions – the northern Adriatic coast, Sardinia and parts of Sicily and Corsica. By 133 BC their authority had spread to Greece, the Iberian Peninsula and Carthaginian North Africa. By AD 14 the entire Mediterranean coastline and the lower Nile Valley were under Roman domination (Proudfoot, 1997a). The Roman economy was predominantly agricultural and the majority of the population was occupied in growing food, except in the large urban areas (Hopkins, 1983). In rural areas and small towns, most produce was consumed locally and the volume of inter-regional trade was small. Even sea travel was expensive. However, as more regions were brought under the hegemony of Rome so more land was used for food production with increasing trade throughout the empire. Wheat was the principal export. It was traded within peninsular Italy but most came from North Africa; it was transported down the Nile to Alexandria and thence across the Mediterranean (Thompson, 1983). By the first century AD North Africa was supplying some 50 000 tonnes of grain per year, and also olive oil (Garnsey, 1983).

From the second century AD the Roman empire gradually began to fragment. From the fourth century the Eastern Empire was ruled from Constantinople, which became the centre of the Byzantine world covering Egypt, the Levant, Asia Minor and Greece, until the Arab conquests from the late seventh century. By AD 650 Egypt and the Levant were Arab. In AD 1453 Constantinople was captured by the Arabs of the Ottoman empire and the Byzantine empire collapsed. Ottoman rule spread through the Balkans and along the eastern and southern shores of the Mediterranean (Proudfoot, 1997b). By the sixteenth century the islands of the Mediterranean and the north African provinces of Cyrenaica, Tripolitania, Tunisia and Algeria had been absorbed into the Ottoman world. The empire was predominantly rural. Agricultural production, much of it in the Tunisian Sahel, the Nile Valley and the Orontes Valley in Syria, was based on dry winter farming of wheat, in other words rain-fed crops. In general productivity was low, both in terms of total yield and on a per capita basis. Peasant holdings were extremely fragmented with little incentive to increase production because of punitive taxation. There was some irrigated agriculture of high-value crops such as tobacco, grapes, sugar, rice and cotton but it was the exception rather than the rule (Proudfoot, 1997b). The Ottomans remained in control of some parts of the eastern Mediterranean until as late as 1920, though their power in eastern Europe and the Mediterranean brought them into a series of conflicts with Russians and other Christian Europeans. During the 1820s they lost power in the Balkans, which had been one of their most productive regions. Loss of food resources from this area and the migration of refugees from the Balkans to Turkey put a severe strain on the Ottoman economy. By the end of the nineteenth century the French had taken power in Algeria and Tunisia. In Egypt there had been an economic boom in the first half of the nineteenth century under the Ottoman governor, Mohamed Ali Pasha, who sought to turn the province into an independent principality. He introduced a cotton monoculture which generated sufficient wealth to create a virtually independent Egypt until 1882, when it became a protectorate and colony of the British.

The western part of the Roman empire from the fourth century AD remained with Rome but it collapsed quite rapidly as a political entity. During the sixth and seventh centuries Islam spread from its heartland in the Middle East. The Islamic religion was founded by Mohammed at Medina in AD 622, the first year of the Islamic calendar (Graham, 1997). By 650 Alexandria in Egypt had become an Islamic capital; by 711 Tunis was founded as an Islamic city and the Arabs quickly took all of the Maghreb. They also crossed the Strait of Gibraltar and moved rapidly through the Iberian Peninsula in less than a decade. Although they crossed the Pyrenees they were defeated by French forces near Poitiers in 732 and retreated across the Pyrenees to consolidate their rule over the Iberian Peninsula. They introduced new crops and farming practices, the environmental impacts of which are considered in Chapter 8. The reconquest by Spanish Christians (the *reconquista*) took some seven centuries in a rather spasmodic process. The barrier between Islam and Christianity has been described as 'permeable', across which influences from both cultures and the Jewish religion intermingled (Graham, 1997). Medieval Spain was perhaps rather more pluralist and less Catholic than the many interpretations of the *reconquista* as a battle between Christianity and Islam suggest. Moorish rule fragmented into a number of smaller states – Córdoba and Seville were captured in 1236 and 1248 respectively, and the kingdom of Granada fell in 1492. The Moors were finally expelled from the Peninsula and Spain became unified by the rule of Ferdinand of Aragon and Isabella of Castile.

Elsewhere in the former western empire of Rome, there are suggestions of a continuity of agricultural practices, land-holdings and rural settlements from classical to medieval times (Delano-Smith, 1979), albeit with the introduction by the Moors of some of the food crops already mentioned – melons, citrus fruits, sugar. The Arabs also introduced refinements in irrigation based on hydraulic technology, but essentially medieval agriculture was still dominated by wheat, vines and olives with transhumant sheep and goat rearing. Gradually, as in parts of southern France, tracts of forest and garrigue were eroded by settlement expansion, but not permanently as previously cleared land became revegetated as it was abandoned. The people of the Middle Ages were threatened by malaria, by plague (which apparently arrived in Europe in 1347 on an Italian merchant ship) and supposedly by environmental deterioration caused by slope erosion and deforestation (Graham, 1997). The lowest ebb of medieval urban and economic life in the western Mediterranean is regarded by some as the ninth century. Throughout this period some trade continued between Mediterranean coastal ports, although there is disagreement as to the extent to which the Arabs controlled the Sea and prevented trade. Not until the eleventh century did trade revive based around the Italian city-states of Venice, Genoa and Pisa. These cities dominated long-distance trade of spices, sugar, dyes, silks and precious stones between Europe and Asia Minor, central Asia and China. Eventually trade also extended westwards to French and Spanish ports and south to North Africa (during the thirteenth century). The wealth generated by this trade helped to support the Renaissance and the overseas expansion of Europe and trade with the New World, leading to a decline in the power of Mediterranean states and the rise of Atlantic Europe and the Baltic ports.

During the twentieth century urban growth in many of the countries of southern Europe was at the expense of rural areas. Change in agricultural practices and abandonment of land have led to increases in forest area. In Mediterranean France, for example, forest cover increased by some 22% between 1965 and 1976. The reasons suggested for this are (Le Houérou, 1981): an abandonment of marginal lands on steep slopes with low productivity, which develop towards garrigue and ultimately a forest cover; the progressive abandonment of grazing; the widespread adoption of a monoculture such as grapes; the extension of irrigated fruit and vegetable crops; the extension of residential zones into rural areas and the growth of second homes; the creation of tourist complexes along the coast; and finally, the decentralisation towards the Mediterranean of high technology firms.

Patterns of twentieth-century vegetation change in southern Europe are in contrast to those of North Africa and Turkey where forest cover is generally declining. The northern and southern shores of the Mediterranean therefore face quite different pressures on the environment. These issues will be brought out further in this book.

1.3 Literature of the Mediterranean world

The literature on the Mediterranean world is vast and varied, and the following paragraphs provide only a brief summary of some of the more important works. The earliest literature dates back as far as any European writings. Thus there are many accounts from ancient Roman and Greek times, some of which are historical, geographical and ecological. Among the earliest of these are the botanical and agricultural researches of Theophrastus of Lesbos. A disciple of Aristotle, Theophrastus lived to the venerable age of 85 and died in either 288–287 or 287–286 BC leaving two major works, *Historia Plantarum* (Enquiry into Plants) and *de Causis Plantarum*. In these he detailed the ecological needs of a large range of plants found in the eastern Mediterranean, both in their wild form and as cultivated crops. Although our scientific understanding of plant ecology has obviously advanced since then, he had explanations for plant habits which many of us would recognise today. Latin authors published similar works later, for example Cato the Elder, 234–149 BC.

Fernand Braudel's (1972, 1973) *The Mediterranean and the Mediterranean World in the Age of Philip II* stands as one of the great works on the Mediterranean. In two volumes, it looks back to the second half of the sixteenth century when Philip II was king of Spain. It is not just a historical account of the Mediterranean region at that time, but is also geographical for he describes the climate, landscape, the rhythms of life and the activities of people from peasants to nobility and so provides a context for the sixteenth-century world. This epic work has been described as a classic to which all Mediterranean scholars must pay homage (King, 1997a).

In contrast to Braudel's focus on the sixteenth century, there are a few volumes which have looked forward to the twenty-first century, addressing the immense impact of people on the region and the possible environmental and economic consequences. The Blue Plan (Grenon and Batisse, 1989), a UNESCO project, established a set of scenarios concerned with the coastal region of the Mediterranean, examining its environmental and resource problems. It was followed by a review of the impact of contemporary global warming on the region in terms of its natural and human environment (Jeftic *et al.*, 1992).

The interaction of people with the Mediterranean landscape has produced a rich field of writing. Some of this has been regionally based, for example the western Mediterranean (Delano-Smith, 1979), or thematic, such as McNeill's (1992) book on the mountains of the Mediterranean and Thirgood's (1981) history of forest resource depletion. Grove and Rackham (2000) provide an ecological history of the European Mediterranean lands. Contemporary geographical issues are addressed by an edited volume on environment and society (King *et al.*, 1997). The potential for severe land-use degradation, perhaps even desertification, is examined by recent volumes such as Fantechi and Margaris (1986), Brandt and Thornes (1996), Mairota *et al.* (1998) and

Conacher and Sala (1998). The last reviews the subject in all five regions with a mediterranean-type climate.

There are a number of research volumes which relate specifically to the ecology of mediterranean-type ecosystems. The starting point for this research, within the framework of the International Biological Programme, was whether very similar physical environments in the five regions of the world with mediterranean-type climates but with phylogenetically dissimilar organisms, would produce structurally and functionally similar ecosystems – in other words whether evolutionary convergence could be recognised (di Castri and Mooney, 1973). This volume was followed by a comprehensive review of mediterranean-type shrublands (di Castri *et al.*, 1981) and later by further studies on the role in ecosystem functioning of nutrients (Kruger *et al.*, 1983), fire (Moreno and Oechel, 1994), biodiversity (Davis and Richardson, 1995), global change (Moreno and Oechel, 1995) and landscape disturbance (Rundel *et al.*, 1998). The fact that mediterranean-type ecosystems are prone to ecological invasions has also been examined in some detail (for example, Groves and di Castri, 1991). Much of the research in these volumes provides a basis for the current book.

Finally, there are a number of books which cover the flora of the Mediterranean region, many with detailed coloured photographs to aid identification of plants and some with essays on their origins. Perhaps the best known is Polunin and Huxley (1987) but there are also Burnie (1995) and Schönfelder and Schönfelder (1990). Some volumes are more regional in their coverage, such as Polunin and Smythies (1973) for southwest Europe and Huxley and Taylor (1977) for Greece and the Aegean.

1.4 Structure of this volume

In attempting to cover the fields of physical geography and human impacts in a range of terrestrial and marine environments, each book in the *Ecogeography* series, of which this volume is a part, follows the same structure. The nature of the physical environment is covered in Chapters 2, 3 and 4 on climate, topography and drainage, and soils, respectively. Chapters 5, 6 and 7 are on plants and animals, communities, and ecosystems. Chapter 8 covers land use and, finally, Chapter 9 considers contemporary environmental issues in the Mediterranean region.

Adopting this structured approach leads to difficult decisions as to where to place material. Inevitably there is a large measure of overlap between much of the ecological subject matter in Chapters 5, 6 and 7. For example, the ecological consequences of fire, which are integral to mediterranean-type ecosystems, are covered in different ways several times. A freer interpretation of the structured themes might have fitted the subject matter rather better; therefore, in an attempt to help the reader, a number of topics and themes which run through this volume are highlighted in Table 1.3 showing where these occur throughout the book. As far as possible detailed material on a particular subject is not repeated. Hence within the text there are also cross-references to other sections of the book which contain allied material.

One of the limitations in writing a text with a very specific location, as opposed to a more general text on ecogeography, is the scope for which there is coverage and debate about any of the philosophical issues germane to the fields of ecology and geography — for example the importance of the ecosystem paradigm within ecology or the replacement of a mechanistic perception of ecosystems by one which relates to their dynamic nature and organisation. It is not intended that this volume should cover these contested issues in detail, though some of these are identified. Instead there are suggestions for further reading at the end of each chapter which should direct readers to other texts that provide background on the topics covered. Generally these do not include books of especial relevance to the Mediterranean or to mediterranean-type ecosystems. References to these are included in the conventional manner within the text.

Given its geographical location, much of the literature on the Mediterranean has been published in Portuguese, Spanish, French, Italian, Greek and Turkish. This is particularly so of the early work on vegetation communities by French researchers until the 1960s. However, the recognition of mediterranean-type ecosystems elsewhere and their similarities in climate and vegetation morphology has led, from the 1960s, to co-operation between scientists across the five regions. Fortunately for those of us in the English-speaking world, this has meant the arrival of a large literature on Mediterranean ecology and physical geography in English, though reflecting some of the earlier work. This is evident in the references at the end of this volume. As far as possible these are to books and journals which are readily accessible and the majority of them are written in English.

One of the problems facing non-specialist readers of ecology books is the profusion of taxonomic names for plants and animals. Unfortunately this is inevitable for many species do not have common names, and when they do these vary from language to language, place to place and sometimes from user to user. For this purpose a note on the Linnaean system of nomenclature of plants and animals is given in Appendix 1. Within the text, common names are used where there is little disagreement as to the appropriate name, but taxonomic names are always given for these in order to avoid confusion. Where there are no common names, taxonomic ones are, of course, the only names used.

The present Mediterranean landscape is, in part, a reflection of its geological evolution and Appendix 2 gives the geological chronology, which stretches over a very long time scale. There is also a wealth of information from the Mediterranean region from more recent times, from archaeological and historical evidence, and many classicists and historians are comfortable with the designation of events as taking place in years BC or AD. For many physical geographers, geologists and Quaternary scientists, however, dates are more usually expressed as BP (Before Present). Unfortunately, converting BC/AD dates to BP dates and vice versa is not merely a matter of adding or subtracting approximately 2000 years. A radiocarbon year differs from a calendar year — the radiocarbon ages of dated samples diverge systematically from their calendar

Table 1.3 Major themes of the book identified by chapter and section

	Chapter							
Theme	Climate (2)	Topography and drainage (3)	Soils (4)	Plants and animals (5)	Communities (6)	Ecosystems (7)	Land use (8)	Environmental issues and conservation (9)
Biodiversity				5.2 Historical development of present Mediterranean flora 5.3 Historical development of Mediterranean animal populations	6.4 Mammal extinctions	7.2 Ecosystem functioning 7.3 Ecosystem development 7.4 Disturbance	8.5 Woodland management 8.7 Grazing	9.1 Plant extinctions 9.3.1 Hotspots conservation 9.3.4a *Dehesa* conservation
Soil erosion		3.3.2 Rock type 3.4 Landscape change	4.3 Soil erosion			7.1.2a Fire	8.8 Intensive agriculture 8.10 Landscape stability	9.2 MEDALUS
Fire			4.5 Soils	5.5 Adaptations of plants	6.2.2 Evergreen oak woodland	7.1.2a Ecosystems 7.2.7 Species richness 7.3.2 Disturbance 7.5 Global change	8.5 Woodland management 8.7 Grazing	9.2 MEGAFiReS

	Ch. 2	Ch. 4	Ch. 5	Ch. 6	Ch. 7	Ch. 8	Ch. 9
Grazing			5.2.6 Invasive plants 5.4.4 Aromatic plants	6.2.4 Steppe and grassland communities	7.2.7 Species richness 7.3 Disturbance	8.5 Woodland management 8.6 Transhumance 8.7 Grazing and browsing	9.3.4a *Dehesa* conservation
Global change	2.8 Climate				7.5 Effects on ecosystems		
Deforestation/clearance		4.3 Agricultural abandonment				8.3 Neolithic clearance Box 8.1 Pollen indicators	
Agricultural abandonment		4.3 Soil erosion				8.8.1 Agricultural intensification	9.3.3 EU policies
Agricultural intensification						8.8 Twentieth-century agriculture	9.3.3 EU policies
Invasive plants			5.2.6 Exotic flora				

age in samples more than about 2500 years old. Radiocarbon dates therefore need to be calibrated to give calendar years. Calibration has resulted in major revisions of archaeological chronologies and therefore of cultural history, especially in places like the Mediterranean where there are contested debates such as the timing of agricultural domestications. Geologists, palaeoecologists and environmental scientists have been less ready to use calibrated dates, preferring to stick with dates expressed in radiocarbon years, as 'so many years BP'. Because much of the literature which uses radiocarbon dating cited in this book comes from geological or palaeoenvironmental studies which use uncalibrated dates, these ages are quoted following the convention of giving uncalibrated dates, expressed as 'years BP'. However, for those readers not familiar with this, Appendix 3 provides information on the calibration of radiocarbon dates and a table for converting dates. A date expressed in Cal. yr BP (calendar years before present) is given this way when the original author only referred to calibrated radiocarbon dates. Dates which are given as BC/AD are those which have a firm basis in the historical record.

1.5 Further reading

King, R., Proudfoot, L. and Smith, B. (1997) *The Mediterranean: Environment and Society*, Arnold, London.

Chapter 2

Climate

Sun, sand and sea – the stimuli for tourism in the Mediterranean. From the mid-nineteenth century to the early twentieth century the French and Italian Rivieras became the winter playground for wealthy northern Europeans. Visitors came for the sea-air, sea-bathing and sun to restore their constitutions. Grand hotels were built overlooking the sea, and life, for some at least, was hedonistic. From the 1920s the emphasis switched from winter to summer vacations as sun-bathing and pursuit of a tan became fashionable. What drew these people to the Mediterranean? In part, the answer must lie with the climate, even if at other times visitors have sought out culture, history and good food. What, therefore, makes a mediterranean climate so distinctive, one which can be recognised in several other areas of the world? In what ways does a mediterranean-type climate influence ecosystem processes, animal and plant life, geomorphological processes and land degradation, to impart a characteristic 'Mediterranean' landscape? This chapter explores the distinctiveness of the climate of the region and, although it might seem contradictory, emphasises its variability, both spatial and temporal.

2.1 General characteristics

The Mediterranean Basin is a region of transition between the sub-oceanic, cool temperate zone of Europe, and the high-pressure, arid subtropics of North Africa; between the pressure systems of the Atlantic, and the monsoon climates of the Indian subcontinent and East Africa. Thus particular patterns of atmospheric general circulation outside the Mediterranean region interact with and determine those within the Basin. Such interactions explain climatic differences from place to place; climate is not uniform across the Basin, even if it is basically consistent, with hot dry summers and mild wet winters. Other general characteristics include strong winds, occasional torrential rains and marked local variations as a result of relief, aspect, altitude and distance from the sea.

Figures 2.1a–2.1c illustrate some of the general climate patterns of the Mediterranean. Summer temperature patterns are not dissimilar across the Basin, but winter temperatures decline northwards and with altitude. Precipitation patterns show a broad contrast between double maxima of autumn and spring rainfall in the north, and a winter maximum in the south. This reflects the

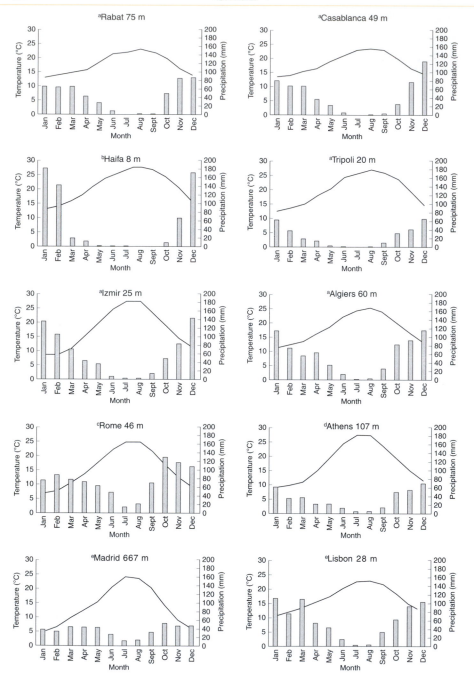

Figure 2.1a Climatic summaries for the Mediterranean region: mean monthly temperature (°C) and total monthly precipitation. Data sources: [a] Griffiths (1972), [b] Taha *et al.* (1981), [c] Cantù (1977), [d] Furlan (1977), [e] Linés Escardó (1970).

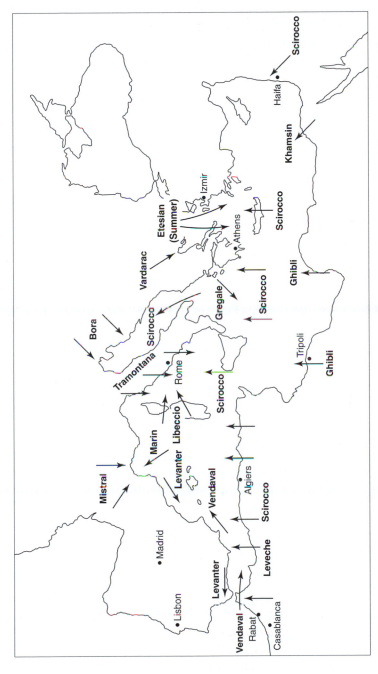

Figure 2.1b Climatic summaries for the Mediterranean region: map of the main regional winds of the Mediterranean. Adapted from HMSO (1962).

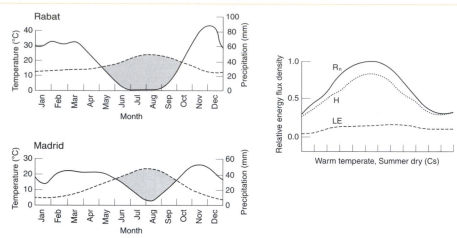

Figure 2.1c Climatic summaries for the Mediterranean region: bioclimatic summaries and annual energy balance for a mediterranean-type climate. After UNESCO/FAO (1963) and Kraus and Alkhalaf (1995).

transition from a continental interior-type climate with a summer maximum, to the mediterranean-type with a winter maximum. Annual precipitation totals are also higher in the western Mediterranean than in the eastern subsection, a consequence of different areas of frontagenesis, or depressional activity.

2.2 Winter rain, summer drought and wind

2.2.1 Winter rain

The wet season in the western Mediterranean is usually established by late October, with the probability of rainfall in any five-day period increasing from 50–70% early in the month to 90% by the end of the month (Barry and Chorley, 1998). The onset of the wet season occurs later towards the south and east of the basin (see Figure 2.1a), which also helps to explain the timing of the wettest months. For example, in Israel this is in December or January (Perry, 1997). The wet, windy and mild weather continues through until about March. Spring occurs gradually and is unpredictable; it may be associated with extreme weather conditions (see Section 2.5).

2.2.1a Cyclonic activity

Winter rainfall is associated with the formation of depressions in the differing regions of the Mediterranean Basin (see Figure 2.2). Because sea-surface temperatures in winter are higher than the mean air temperature, by about 2 °C in January, incursions of cooler, maritime polar air from the Atlantic or continental polar air from southern Europe are warmed over the Mediterranean Sea (Barry and Chorley, 1998). This leads to convective instability producing frontal precipitation. Atlantic depressions entering the western Mediterranean account

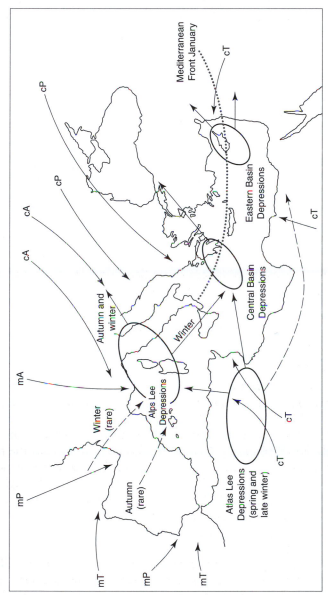

Key: mT, Maritime tropical air mass; mP, Maritime polar air mass; mA, Maritime arctic air mass; cT, Continental tropical air mass; cP, Continental polar air mass; cA, Continental arctic air mass

Figure 2.2 Depressional systems of the Mediterranean Basin. After HMSO (1962) and Barry and Chorley (1998).

for 9% of depressions affecting the region. A further 17% originate south of the Atlas Mountains, especially in late winter. The warming of the Atlantic maritime polar air produces a 'mediterranean' air mass, the southern boundary of which forms the Mediterranean Front, located over the Mediterranean and Caspian Seas in winter (shown on Figure 2.2 and later on Figure 2.3(b)).

Depressions also develop within the Mediterranean Basin itself, accounting for 74% of the region's low-pressure systems. In the western Mediterranean Basin to the lee of the Alps and the Pyrenees, there are the Genoa-type depressions associated with significant instability and intense precipitation along the warm front, and heavy showers and thunderstorm activity behind the cold front. These depressions only occasionally move eastwards far enough to affect the eastern Basin (Wigley and Farmer, 1982) but their trajectories are influenced by relief effects. For the eastern Basin, depressions forming in the central Basin and eastern Basin ('Cyprus Lows') are most important. Occasionally their origins may be Genoa-type or Atlas Mountain depressions. Both central Basin and eastern Basin depressions may move northeastwards or eastwards. Regeneration of these depressions may occur when they meet incursions of cold continental polar air from southeastern Europe and Russia.

2.2.1b Anticyclonic conditions

Winter weather is not solely dominated by low-pressure systems but is highly variable. The depressional pathways are influenced by the position of the Subtropical Westerly Jet Stream, and this is highly mobile. Depressions may pass to the north of the Mediterranean Basin, resulting in anticyclonic conditions over the Basin itself and therefore stable, dry weather. These are dominant 25% of the time over the Basin as a whole, and for 48% of the time in the western Basin (Barry and Chorley, 1998). As a result rainfall is not evenly distributed through the winter, but concentrated into a few days per month or season. This is important when considering the need for water by plant or animal life in the region (see Section 5.4 for greater discussion on this subject).

2.2.2 Summer drought

Summer drought results from the expansion of the Azores high-pressure system into the region, bringing stable conditions with hot, dry weather. There may be some rain as depressions can develop, but they are rarely strong, and air mass contrasts are far weaker than in winter. The duration of the drought is greater in the southern and eastern areas of the Basin, extending for up to five consecutive months. Occasionally intense heatwaves occur, as for example in Greece during 1987 and 1988. As well as considerable thermal stress and discomfort, this led to an increase in forest fires (Giles and Balfoutis, 1990). From the biological point of view drought may be exacerbated by desiccating regional winds of continental tropical origin, such as the khamsin of Egypt and the sirocco winds of Algeria and the Levant (see Section 2.2.3). Occasionally, monsoon air masses may bring summer rain to the eastern Mediterranean (Wigley and Farmer, 1982) and the occurrence of this type of rain may have been more frequent in earlier times. The influence of the monsoon is discussed further in Section 2.4.2.

2.2.3 Wind patterns

Strong winds are also a significant feature of Mediterranean weather patterns. In winter these are associated with depressions but are also influenced by topography. For example, the mistral is a cold, northerly wind in the Gulf of Lion, which develops when a Genoa-type depression forms east of the Azores high-pressure ridge. The wind is then funnelled along the Rhône Valley. The mistral may last for several days, and is at its most intense between December and April, although it also blows in summer (Barry and Chorley, 1998). The mistral increases the risk of fire (Wrathall, 1985) and so can be an influential factor in vegetation communities. Similar winds occur in other localities (see Figure 2.1b), such as the bora in the northern Adriatic, and the northerly or northeasterly tramontana on the Catalan coast of Spain, which also affects the Balearic Islands, northern Corsica and the west coast of Italy. It blows in advance of an anticyclone following the passage of depressions eastwards through the basin, especially in winter and early spring. As well as extreme winds, frost and snow can occur within the region.

In summer desiccating regional winds may blow, such as the khamsin of Egypt, the leveche of southeastern Spain, and the scirocco winds of Algeria and the Levant (see Figure 2.1b). These are warm, south or southeasterly winds blowing in advance of depressions moving eastwards through the Mediterranean or across North Africa. With a Saharan origin, these are dry and dusty and may deposit dust in Malta, Sicily, southern Italy and Greece, and indeed there is speculation about the importance of this dust contribution, over long periods of time, to soil formation in southern Europe (see Section 4.1.3). As the scirocco winds cross the sea, picking up moisture, they reach southern Europe no longer dry, but hot and humid. As well as the many local winds, in coastal regions land and sea breezes are important.

2.3 General circulation – the Mediterranean region _____

The climate patterns of the Mediterranean region, and for that matter any world region, can be partly explained by general circulation patterns. Surface climate conditions are closely related to upper-air, mid-tropospheric, circulation. At temperate latitudes in the northern hemisphere, the upper-air flow is a meandering westerly flow maintained by the Earth's rotation and temperature differences between equatorial and polar regions. Surface topography, the distributions of land and sea, sea-ice and snow cover influence the meander patterns, which alternate between the ridges and troughs which make up the long waves of the upper-air flow (Barry and Chorley, 1998). The velocity of this westerly flow may reach 45–67 m/s and maximum speeds are often found at about 30° latitude (over North Africa) and at an altitude of between 9000 and 15 000 m. This is the jet stream. There are three component jet streams

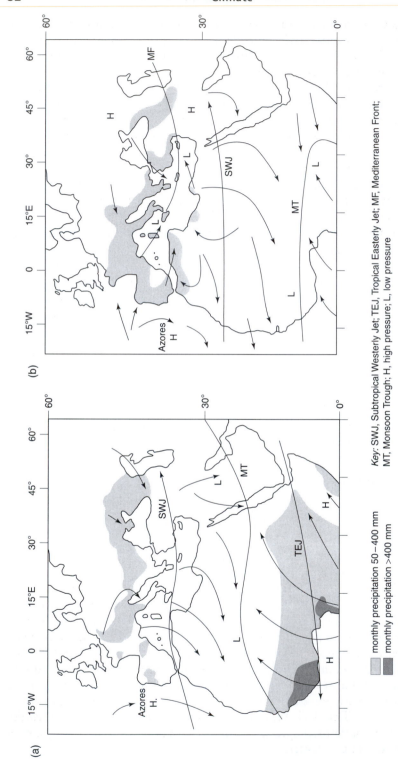

Figure 2.3 Distribution of surface pressure, winds and precipitation for the Mediterranean region. (a) July and (b) January. Adapted from Barry and Chorley (1998).

Key: SWJ, Subtropical Westerly Jet; TEJ, Tropical Easterly Jet; MF, Mediterranean Front; MT, Monsoon Trough; H, high pressure; L, low pressure

monthly precipitation 50–400 mm

monthly precipitation >400 mm

which affect the Mediterranean region and these are shown in Figure 2.3: the Polar Front Jet Stream, the Subtropical Westerly Jet Stream and the Tropical Easterly Jet Stream. Of these, the Subtropical Jet Stream is more persistent and its location helps to determine the location of the mean jet stream. Although the Polar Front Jet Stream and Subtropical Westerly Jet Stream are well separated over monthly or longer time scales, they may merge on a daily basis. The southward excursion of the Polar Front Jet Stream may be longitudinally aligned with the northward excursion of the Subtropical Westerly Jet Stream, especially over the Black Sea region. Although a weak feature of jet stream circulation, it may bring upper tropical air into mid-latitudes in winter time, and this may have occurred more frequently in the past (Wigley and Farmer, 1982).

The location of the jet streams varies seasonally, being further north in the northern hemisphere summer (see Figure 2.3(a)). The location of the Subtropical Jet Stream also determines the location of surface pressure conditions, notably the subtropical anticyclones at about 30° latitude, although the relationship between the two is complex. For the Mediterranean region, the most important subtropical high-pressure cell is that of the Azores. Like other subtropical anticylones, this moves a few degrees polewards in summer and equatorwards in winter, depending on the meridional (north–south) temperature difference between low and high latitudes. The mean position of the jet streams is highly variable, but changes in surface climate can be interpreted in terms of shifts in these positions, both today and for past climatic fluctuations.

2.4 Secondary influences on climate

In addition to general circulation patterns other factors influence the climate of any location, for example topography.

2.4.1 The influence of topography

Rainfall in the Mediterranean region is not just the result of cyclonic or convective activity; the relief or orographic component is also important. Given the localised variability in relief, precipitation may vary considerably over very short distances. For example, Navacerrada, in Spain, is only 50 km from Madrid, but at an elevation of 1860 m, its rainfall is approximately 1300 mm, or three times that of Madrid, with about 430 mm (Linés Escardó, 1970). Figure 2.4 illustrates mean annual rainfall across the Iberian Peninsula for the period 1931–1960. It is highly irregular, reflecting a complex relief pattern. In some areas, precipitation exceeds 1500 mm per annum, such as in the Serra da Estrêla, the western central Pyrenees, the Sierra de Gredos (Central Range) and the Sierra de Grazalema (Cádiz). In general, these are areas with elevations greater than 1500 m. In contrast, in the Ebro Valley precipitation is as low as 300 mm per annum, and near Almería, below 200 mm per annum, the lowest total in Europe. This is a low-lying, rainshadow location (Linés Escardó, 1970). Similar patterns of precipitation with respect to relief are found in other Mediterranean countries.

Figure 2.4 Precipitation map of the Iberian Peninsula. After Linés Escardó (1970).

2.4.2 Extension of the monsoon

In addition to the westerly upper-air flow described in Section 2.3, the eastern Mediterranean and Middle East may be influenced by the Indian monsoon system (Wigley and Farmer, 1982). Although an over-simplification, in summer strong heating of mid-tropospheric air over the Tibetan plateau produces an upper-air anticyclone. While this weakens the Subtropical Jet Stream south of the Himalayas, it strengthens the Polar Front Jet Stream to the north and a strong easterly jet forms, the Tropical Easterly Jet Stream. The prevailing westerly winds migrate northwards of the Himalayas and moist tropical maritime air crosses the Indian subcontinent. As the monsoon advances, the easterly jet strengthens and extends over India, southern Arabia and into East Africa, and so over the headwaters of the River Nile, between July and September (see Figure 2.3(a)). This may bring rain to the headwaters of the Nile, leading to increased discharges. Increased precipitation may also occur in southern Arabia, though it is unlikely that monsoon rain would penetrate across Afghanistan and into the Middle East.

Precipitation in the upper Nile, and therefore the unregulated flow of the Nile into the Mediterranean, is directly related to the intensity of the easterly jet stream. Persistent high Nile discharges are thought to have been responsible for the formation of sapropels – thick layers of black mud rich in organic matter – in the Mediterranean Sea off the Nile delta (see Section 3.5.1). The two most recent sapropels are dated to the transitional period at the end of the last glacial stage and beginning of the Holocene, a time of intensified and northward displacement of the African monsoon over the Ethiopian highlands.

2.5 Extreme weather conditions and short-term climate variability

The generalisations about Mediterranean climate outlined above mask a diversity of climates and weather conditions. Short-term climatic oscillations are apparent in meteorological records, as can be seen by fluctuations in temperature and precipitation in the Mediterranean during the twentieth century. Climatic variability is also high from year to year and between coastal, lowland and upland localities.

2.5.1 Extreme weather

The transition from summer to winter is the time of greatest variability in weather, such as heavy rainfall leading to flooding in North Africa in September 1969, in southeast Spain in October 1973, and in Piedmont and Liguria (Italy) in October 1977. At this time of year the difference between air and sea-surface temperatures is at its maximum and therefore potential instability is at its greatest (Perry, 1981). An excellent example of a severe storm at this time of year comes from Spain and the Balearic Islands in September 1989. It was not uniform in its severity of impact across the islands and the distribution of precipitation was unlike that of 'normal' rainfall events, being highly localised (Wheeler, 1991). On Mallorca heavy rain, exceeding the 150 mm 100-year return period of 24-hour rainfall, fell along the southeast coast of the island. Generally the wettest part of the island is along the northwest coast, but in September 1989 that area was relatively dry. The severe weather resulted from onshore winds along most of Spain's southeasterly coast and the Balearic Islands. Their origin was North Africa and so they were hot and dry, but picked up large amounts of water vapour crossing the Mediterranean Sea. On Mallorca the coastal relief, though not high, was sufficient to initiate uplift and realise the potential instability, resulting in heavy downpours. Severe flooding closed roads, and destroyed buildings and crops, as well as leading to the deaths of at least three people.

While it rarely snows in the narrow coastal belts of the Mediterranean region, snowfall and snow cover are important in mountainous interiors, such as the Central Apennines of Italy. Here there is intense snowfall because of moisture-rich maritime air, overlying cold air originating in the Balkans or Russia (Cantù, 1977). Snowmelt is an important contribution to drainage basin hydrology (see Chapter 3) and to water resources in the Mediterranean Basin, both for semi-natural vegetation and for agricultural productivity.

In general coastal regions are frost-free, but altitude and distance from the sea increase the risk of frost. This is of importance to agricultural production. For example, citrus crops require a frost-free climate, and grow best on level ground, preferably at a slight elevation with good air drainage to avoid frost pockets (Mabberley and Placito, 1993).

2.5.2 Short-term climatic oscillations – twentieth-century change

Fluctuations in temperature and precipitation in the Mediterranean region during the twentieth century are quite marked. While there may be little change in annual totals between one thirty-year period and another, as in Portugal between 1932–60 and 1961–90, seasonal distribution has altered considerably. Figure 2.5 shows a sharp reduction in rainfall in March for the period 1961–90 compared with 1932–60. Over the same periods rainfall increased in October, November, January and February, leading to a slight increase in autumn and winter precipitation (Corte-Real *et al.*, 1998). In effect, spring-time in Portugal ends earlier and is becoming drier and thus the onset of summer and drought occurs earlier. There is also evidence that rainfall totals during the period 1951–89 decreased in many areas of the Mediterranean Basin, especially the northern shores. The reduction in rainfall is statistically correlated with increased frequency and persistence of anticyclonic systems over the western-central Mediterranean, particularly the Azores High. Plotting the elevation of the 500 hPa pressure field (Figure 2.6(a)) for the period 1946–1989 for Algiers shows a steady increase in height. This trend was representative of the whole western-central Basin and led to endemic drought through the 1980s, culminating in severe drought from September

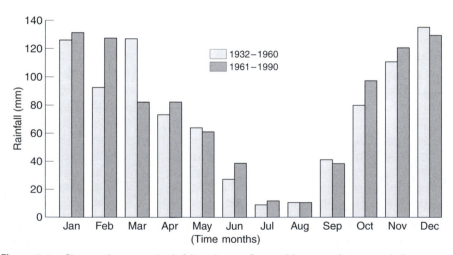

Figure 2.5 Changes in seasonal rainfall cycle over Portugal between the two periods 1932–1960 and 1961–1990. After Corte-Real *et al.* (1998).

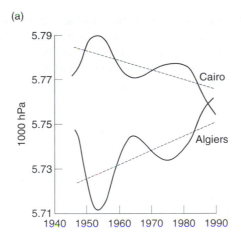

(a)

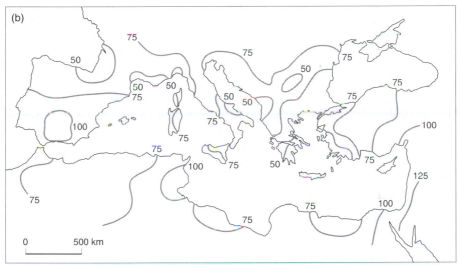

(b)

Figure 2.6   (a) Change in the height of the 500 hPa level over Algiers and Cairo for the period 1946–1989. After Corte-Real *et al.* (1998). (b) Percentage of normal (1951–1980) rainfall observed from September 1988 to April 1990. After Corte-Real *et al.* (1998).

1988 to April 1990. The percentage of normal (1951–80) rainfall observed for that period is shown in Figure 2.6(b). Areas with around 50% of the normal rainfall are common along the coastal areas of the northern Mediterranean. Such a drought means a longer period of biological stress, affecting both semi-natural vegetation communities and crops, and could lead to a reduction in vegetation cover and increased risk of soil erosion. There would also be the need for longer storage time of reservoir water for irrigation, domestic and industrial use. The pattern of the western-central Mediterranean has not been repeated across the entire Mediterranean; climate anomalies may be quite different between the western and eastern, or northern and southern sectors of the Basin. This is evident in Figure 2.6(a), which shows that the elevation of the 500 hPa pressure field for Cairo, typical of the eastern Mediterranean, was opposite to that of Algiers (Corte-Real *et al.*, 1998).

While near-surface air temperatures averaged over the globe have increased by about 0.5 °C since the mid-nineteenth century (Wigley, 1992; Houghton *et al.*, 1996) the warming trend has not been consistent through time and space. For example, during the period 1950 to approximately 1990, there was a noticeable cooling in the whole of the Mediterranean Basin (Wigley, 1992). This also occurred in much of the northern hemisphere from the mid-1940s to early 1970s (Barry and Chorley, 1998). However, within the period 1950–1990, the patterns were not consistent across the Mediterranean Basin. For the period 1967–1986 there was cooling in the eastern Basin, while temperature trends in southwestern Europe (the Iberian Peninsula, south and central France) showed a warming in the 1980s equivalent to about 2 °C per century (Beniston and Tol, 1998).

2.6 Establishment of a mediterranean–type climate in the Mediterranean Basin

A recognisable 'modern' mediterranean-type climate in the Mediterranean region, with a rhythm of summer drought and winter rainfall, first appeared approximately 3.2 million years ago during the mid- to late Pliocene. This was the result of a global cooling, the establishment of permanent ice in the Arctic and the intensification of the North Atlantic Drift (Suc, 1984). The bi-seasonality of climate has not been consistent since that time, with major fluctuations in response to changes in Quaternary ice volume. Evidence from lake-level variations around the Mediterranean Basin – southern Europe, the Middle East and North Africa – indicates high levels at the time of the Last Glacial Maximum, about 18 000 years ago, suggesting a wetter climate than today (Harrison and Digerfeldt, 1993). This is at odds with the evidence from pollen records, which indicates *Artemisia*-dominated communities consistent with aridity (see Chapter 5). This apparent contradiction is explained in terms of changed seasonal precipitation patterns. At the time of the Last Glacial Maximum simulations of Mediterranean climate, based on general circulation models, suggest winter temperatures would have been 5–10 °C lower than at present, and 1–3 °C lower in summer (Kutzbach and Guetter, 1986). Simulations also show that over the continental ice sheets of northern Europe strong anticyclonic conditions would have built up, and the upper westerly winds would have been displaced southwards from their present tracks, with surface westerly winds much cooler, having crossed colder or frozen oceans. Winds would have been cold and dry. In summer it is suggested that the Mediterranean region would have been cooler and cloudier, though with no more precipitation. Cooling in both seasons, wetter winters and cloudier summers and so less evapotranspiration would help to explain high lake levels (Harrison and Digerfeldt, 1993). In addition it has been suggested that southward displacement of the jet streams might have increased storm frequency.

This, together with increased snow cover in mountain regions, might have led to greater runoff (Macklin *et al.*, 1995). The presence of *Artemisia*-dominated steppe is thus explained by the combination of low annual precipitation totals and cold dry winters, conditions quite different from today.

2.7 Late glacial and Holocene climate changes

After the Last Glacial Maximum climate fluctuations around the Mediterranean Basin were not synchronous (Harrison and Digerfeldt, 1993). In the Iberian Peninsula the climate became drier around 16 000 BP according to lake-level evidence, with maximum late-glacial aridity between 16 000 and 15 000 BP. This was followed by wetter conditions (compared with both earlier and later periods) in the western Mediterranean from about 12 000 BP – the time of highest lake levels; in Algeria several lakes show high water levels during the early to mid-Holocene. Aridity abruptly recurred around 5000 BP. The pattern in the eastern Mediterranean is different. In the Balkans maximum aridity is dated to between 13 000 and 11 000 BP, though at Lake Konya, in Turkey, water levels were higher between 12 000 and 11 000 BP. In the east, lake levels suggest that there was only a gradual return to wetter conditions, which were at their maximum around 6000 BP. The transition to contemporary drier conditions then took place gradually from about 5000 BP.

The climatic interpretation of the lake-level data for the Holocene is complemented by that from pollen evidence. Huntley and Birks (1983) noted the westward expansion of sclerophyllous vegetation (including evergreen *Quercus* (oak), *Olea europaea* (olive), *Phillyrea* and *Pistacia*) at the expense of deciduous forest and this is considered to be a response to the increasing aridity of the latter half of the Holocene, during which pollen evidence suggests increasing summer drought as the Holocene progressed. In consequence, for the western Mediterranean the present period is the driest phase of the last 30 000 years.

The climatic forcing mechanisms which resulted in these patterns of lake-level and vegetation changes are not yet fully understood. The marked differences between the early and late Holocene climate of the Mediterranean do not accord with general circulation model simulations based on orbitally induced insolation changes and concomitant changes in atmospheric circulation (Harrison and Digerfeldt, 1993). These simulations suggest that there should have been a consistent, strengthened subtropical high-pressure cell maintaining offshore winds, similar to those of present-day summer conditions, over the Mediterranean region throughout the Holocene. They do not predict the early to mid-Holocene high lake levels found in some regions of the Basin. Instead there have been suggestions that the regional Holocene differences in climate are better explained by changes in precipitation and storm tracks across the southern Mediterranean, as a result of gradual changes in the number and location of meanders in the jet steam. Warmer sea-surface temperatures in the North Atlantic and Mediterranean would have increased atmospheric humidity, precipitation and more frequent incursions of tropical air masses into the region than at other times (Macklin *et al.*, 1995). The present aridity may result from the northwards displacement away from the Mediterranean of moist, westerly winds.

The biological significance of these climate fluctuations is that climate stability cannot be considered the norm. Vegetation and animal communities and ecosystem functioning adapt to such changing conditions. In addition, climate fluctuations influence hydrological regimes, soil processes and fire frequencies. These in turn influence vegetation and animal communities. Climatic fluctuations over the past few thousand years have also occurred at the same time as increased anthropogenic activity in the Mediterranean Basin. People are a major determinant of ecosystem processes. It is therefore difficult to establish the extent to which vegetation change, as revealed by palaeoecological indicators, reflects climate change or cultural activity. Set against this uncertainty about the past response of ecosystems to environmental change is the debate about the effects of contemporary and future climate changes and the scale of human impact into the twenty-first century. The issue of future climates is the subject of the remainder of this chapter.

2.8 Future climates

Disparities in climate fluctuations across the Mediterranean, within both the recent past and the longer-term Holocene, should lead to caution in predicting future climate changes across the region as a whole. Nevertheless the Mediterranean region with its marginal climate, being transitional between northern Europe and Africa, is one where the impacts of climate change may be particularly severe (Corte-Real *et al.*, 1998). Modelling future climates is extremely difficult, especially in the Mediterranean where climates are determined by contrasts between land and sea, by local effects, such as topography, as well as by the larger-scale features of atmospheric circulation, such as the location of jet streams and the Azores high-pressure cell. The complexity is further compounded by the nature of General Circulation Models (GCMs) designed to simulate global climates – of the past, present and future. These are complex three-dimensional computer-based models of coupled atmospheric and oceanic circulation. Several such models exist, usually using pre-industrial atmospheric CO_2 concentrations or doubled CO_2 concentrations to produce scenarios for future climate change. While it is not universally accepted that contemporary global warming, nor future warming, is a consequence of increases in greenhouse gases (see Chambers, 1999 for a review of alternative ideas), the output of these models can inform the debate about potential future climates and the effects of climate change on ecosystems and society. Unfortunately there are severe limitations in the use of GCMs at the scale of the Mediterranean given their coarse spatial resolution, as well as in the running of the models themselves. Examination of the limitations associated with the use of GCMs is beyond the scope of this book but good background on the subject can be found in Houghton (1997). However, it is important to recognise that 'their outputs should be considered only as possible scenarios for future climatic change rather than predictions' and that a scenario 'is intended to be an internally consistent picture of possible future climatic conditions' (Wigley, 1992, p. 23).

Despite his recognition of the limitations of global climate modelling, Wigley (1992) describes the projections of four models for the Mediterranean Basin and stresses that the results are 'of extremely dubious quality' (p. 40). Nevertheless, summarising the results of these models shows that 'at best guess' there would be no major precipitation changes. Increases are marginally more likely in the northern part of the Basin with some evidence of a slight decrease in the southern part of the Basin. Although these model outputs indicated seasonal differences around the Basin, there were few regions where the probability of increase or decrease in precipitation exceeded 90%, the sort of result that could occur purely by chance.

Techniques to develop regional-scale scenarios were included in the MEDALUS research programme (see Chapter 4). Even though the spatial resolution was higher in these results compared with those of GCMs, the geographical/locational representation of areas was not entirely realistic and so the model results again need to be treated with caution. Goodess *et al.* (1998) report results from four GCMs combined to give a mean prediction for temperature change, albeit a very simple pattern of temperature change. These suggest greater warming on the peripheries of the Mediterranean Basin – to the northeast and southwest – with least warming over the central part of the Basin – the Sea itself (see Figure 2.7a). More complex results can be achieved by explaining observed temperature variations at a network of meteorological stations in terms of regionally-averaged climate variables such as temperature, sea-level pressure and precipitation, using regression models. The GCM results can then be substituted for the regionally-averaged variables to estimate temperature changes. The resulting map (see Figure 2.7b) thus takes account of geographical variables such as altitude and proximity to the sea. In theory this produces a more realistic scenario, although such modelling still has major limitations. Nevertheless results such as these can help to inform 'thought experiments' as to the effects of climate change on Mediterranean ecosystems.

2.9 Climate and life

Even a cursory glance at maps showing distributions of climate types and biomes reveals a strong correlation. Ecologists and biogeographers have attempted to bring these two together by defining bioclimates, based on the climatic parameters recognised to be important to the physiology of plants and animals. This topic is the subject of Box 2.1, but it also underlies much of the subject matter of the book – the distinctiveness, or not, of mediterranean-type ecosystems and their plants and animals. This chapter has therefore set out to explain the nature of the climate in the Mediterranean region and its role in determining topography and drainage (the subject of Chapter 3), soils (Chapter 4) and its influence on plant and animal life (Chapters 5, 6 and 7). Climate has also played a part in the land-use history of the Mediterranean region (Chapter 8) and contemporary and future climatic changes contribute to the environmental concerns of the region (Chapter 9).

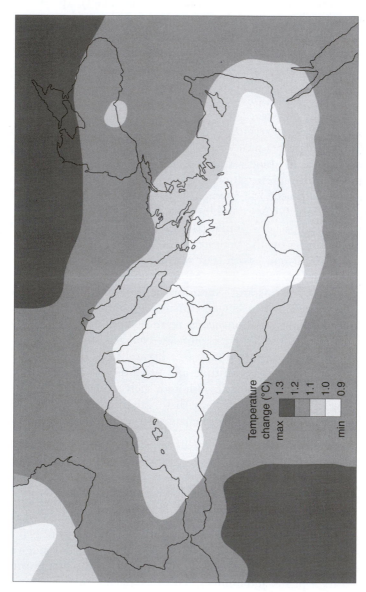

Figure 2.7a Annual scenario of the temperature (°C) change per °C change in global mean temperature. After Goodess *et al.* (1998).

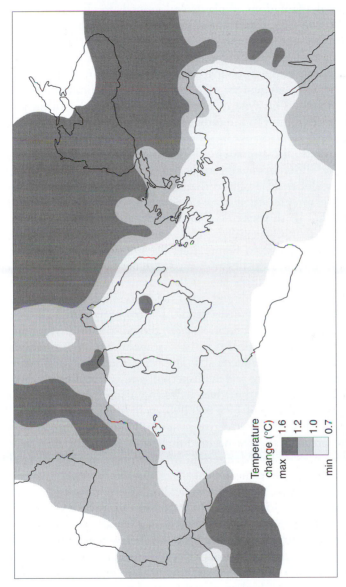

Figure 2.7b   Annual sub-grid scale scenario of the temperature change (°C) per °C change in global mean temperature. After Goodess *et al.* (1998).

Box 2.1   Bioclimates

The combination of hot dry summers and mild damp winters produces an effective physiological drought to which plants respond in different ways. While the physiological adaptations of plants and animals to climate are considered in Chapter 5, it is useful to examine how climatic factors of special importance for living creatures can be mapped to show regions of greatest physiological stress. In the case of the Mediterranean the stress is lack of water. Maps like these are bioclimatic maps. Such maps fall into a tradition of climate classification – the efficient arrangement of information in a simplified form (Barry and Chorley, 1998). This type of research was fashionable within geography and its cognate disciplines in the early to mid-twentieth century in an attempt to produce global-scale maps, in part for examining agricultural potential. This was a reductionist approach to the subject of climate and vegetation response. Several different classifications exist, not all of which are bioclimatic in nature. At a global scale one of the best-known schemes to relate climate parameters to plant growth is that of Köppen, based on temperature and aridity criteria. Under this scheme, a mediterranean-type climate is designated by the letters Csa or Csb, and the characteristic annual energy balance for this climate type is shown in Figure 2.1c. Such a classification can be useful in comparing the characteristics of a mediterranean-type with other climate types.

For the Mediterranean itself and other mediterranean-type climate regions, UNESCO-FAO (1963) published a bioclimatic map predicated on the importance of aridity. The climatic parameters were chosen specifically for that climate type, as opposed to other global types. Inevitably there are limitations in the reliability of such a map, produced at a time when meteorological records were of shorter duration and were widely dispersed, and where climatic boundaries were necessarily arbitrary rather than reflecting a true continuum of climate and were mapped at a scale of 1 : 5 000 000, so hiding local variability. Nevertheless, such a map is still instructive in identifying at the broad scale those regions where greatest stresses occur. Furthermore, such a map reveals the widespread occurrence of a 'Mediterranean' bioclimate in regions well beyond the borders of the Mediterranean Sea.

In the UNESCO/FAO (1963) classification, a dry month is one in which total precipitation (mm) is equal to or less than twice the mean temperature (°C) of that month, $P \leqslant 2T$. When plotted on a climate graph the dry period is evident when the curve for precipitation falls below that of temperature (see Figure 2.1c). The intensity of drought is reflected in a xerothermic, or hot weather drought, index. This is the number of days in a month or year which is deemed dry from the biological point of view. It takes account of the distribution of rainfall, mist and dew through the dry period. From the biological stance the total precipitation figure for a month is an unhelpful statistic,

(continued)

(continued)

for the rain may be concentrated in one day and therefore 'unavailable' to plants during the rest of the month, or may be spread evenly through the month and therefore of more use. In the absence of rain, mist and dew may contribute significant amounts of water. Using the xerothermic index and temperature the 'true' Mediterranean, or 'eumediterranean', bioclimate region is subdivided into four categories: xerothermomediterranean regions which border desert and sub-desert and where drought is intense (between 150 and 200 days per year), thermomediterranean with relatively intense drought (100 to 150 days) and either longer or shorter dry seasons, mesomediterranean, with less intense drought (40 to 100 days), and sub-mediterranean which may have a semi-dry season. Bordering climates with no dry season are termed axeric. Examination of Figure 1.1a shows that a mediterranean-type climate extends east into northern Iran and Iraq, as well as to the southern shores of the Caspian Sea, areas not traditionally considered as 'Mediterranean'.

A common French bioclimatic classification, derived specifically for the Mediterranean region, is that of Emberger (1930) based on a relationship between annual rainfall and the average maximum temperature of the hottest month and the average minimum of the coldest month. This gives six categories based on rainfall: perarid (or very arid), arid, semiarid, subhumid, humid and perhumid (or superhumid) – see Figure 1.1a. Within each there are thermal variants: very hot, hot, temperate, cool, cold, very cold, extremely cold and icy. Temperature variations are determined largely by proximity to desert environments and altitude. Other climatic or climate-associated factors could be considered in deriving bioclimatic indices and maps, such as snow cover, wind, exposure, potential evapotranspiration and soil moisture budget. As soil moisture varies according to soil type and vegetation cover, so evaporation rates will fluctuate depending on actual soil moisture content as well as with temperature and rainfall. A major example of this type of classification was that of Thornthwaite from the 1940s. Although applied to many climate regions, it is not very satisfactory for semiarid areas (Barry and Chorley, 1998).

Multivariate statistics have been used increasingly to delimit regional climatic variability, for example principal components analysis of Mediterranean rainfall and cluster analysis and principal components analysis using monthly precipitation totals and mean temperature for 444 climatic stations (Mazzoleni *et al.*, 1992). The aim of these studies has been to verify some of the more subjective classifications of mediterranean-type climates of the 1950s and 1960s and to recognise objectively distinct sub-regions within the Mediterranean. It also avoids the production of synthetic indices used in delimiting bioclimatic maps. The ultimate regional classifications are, of course, dependent on which climatic parameters are used and are therefore difficult to compare one with another. These attempts also largely reflect a non-English-language, European philosophy of climate classification.

(continued)

(continued)

Bioclimatic classifications have been used, with varying degrees of success, to derive vegetation zones and maps such as the UNESCO-FAO (1970) vegetation map of the Mediterranean zone and the detailed vegetation zoning of Le Houérou (for example, 1990) and Quézel (for example, Quézel and Barbero, 1982). Defining climatically derived vegetation communities is examined further in Chapter 6.

2.10 Further reading

Barry, R.G. and Chorley, R.J. (1998) *Atmosphere, Weather and Climate*, 7th edition, Routledge, London.

Houghton, J. (1997) *Global Warming: The Complete Briefing*, 2nd edition, Cambridge University Press, Cambridge.

Watson, R.T., Zinyowera, M.C., Moss, R.H. and Dokken, D. (1998) *The Regional Impacts of Climatic Change: An Assessment of Vulnerability*, Cambridge University Press, Cambridge.

Chapter 3

Topography and drainage

'The chestnut trees of Anavryti were far below, and as we climbed the steep mountainside and the sun climbed the sky, vast extents of the Morea spread below us. The going grew quickly steeper and the path corkscrewed at last into a Grimm-like and Gothic forest of conifers where we were forever slipping backwards on loose stones and pine-needles. Emerging, we could look back over range after range of the Peloponnesian mountains – Parnon, Maenalus, even a few far away and dizzy crags of Killini and Erymanthus, and, here and there, between gaps of the Spartan and Arcadian sierras, blue far-away triangles of the Aegean and the gulf of Argos. But ahead we were faced by an unattractively Alpine wall of mineral: pale grey shale and scree made yet more hideous by a scattered plague of stunted Christmas trees. These torturing hours of ascent seemed as though they could never end . . . Finally the toy German trees petered out and the terrible slope flattened into a smooth green lawn scattered with flowers and adorned by a single cistus clump with a flower like a sweet-smelling dog-rose.'

Leigh Fermor (1958, p. 13)

Patrick Leigh Fermor's description of climbing the mountains of the Mani in the southern Peloponnese could be typical of many other regions of Greece and the Mediterranean in general. Landscapes are mountainous, of bare rock or with a thin cover of soil, and elevations reach 4000 m (Plate 3.1). Vegetation communities change with altitude, soil and rock type. Rivers flow through steep valleys and gorges down to narrow coastal plains and small isolated lowland areas, separated by hills or mountains. River flow is often ephemeral with discharge strongly controlled by climate – higher in winter than in summer. The coastline is highly dissected and hundreds of islands are found within the Basin, ranging in size from Cyprus to a few hectares. The result is considerable landscape variability, moulded by geomorphological processes, some extending far back in time – the consequence of major climatic changes and long-term tectonic deformations – but also continuing today; earthquake activity is ever present in many regions.

Superimposed on natural processes are those induced by people living in the region – a succession of cultures with differing origins, ethnicity, social

Plate 3.1   The Helikon Massif (>1500 m), Boeotia, central Greece. Note the changing vegetation communities with altitude – coniferous woodlands in the foreground and grass meadows beyond.

complexities and technological attainments. Some areas have been occupied for at least the past 100 000 years and human influence has increased dramatically through this period. The past 10 000 years have seen the spread of agriculture and the emergence of the great Mediterranean civilisations (as outlined in Chapter 1) which have left their mark on the landscape. In the recent past agricultural intensification, and especially the influence of the European Union's agricultural policies, has been very pronounced. Repeated periods of over-intensive land use have led to episodes of vegetation clearance, soil erosion and deposition. Many regions have also been subjected to ethnic and religious conflict, international wars and changing frontiers. This remains an issue today. All of these factors have left their imprint on the landscape and the linkages between anthropogenic activity, geomorphological processes and the physical environment are the subject of this chapter.

3.1 Geological evolution

3.1.1 Tectonic activity

Analysis of surface earthquake activity reveals the role of tectonic activity in unravelling the complex geological evolution of the Mediterranean Basin. Just as climatically the Mediterranean is transitional between Europe and Africa, so it is also a boundary zone between the Eurasian, African and Arabian plates, but with a number of smaller secondary or microplates (Figure 3.1). The

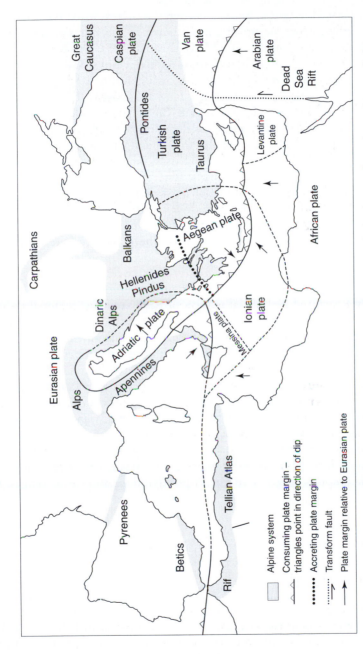

Figure 3.1 Schematic outline of the Alpine system of Europe with present lithospheric plate configuration and types of plate boundary in the Mediterranean Basin. Adapted from Smith and Woodcock (1982) and Macklin et al. (1995).

movement of these plates has produced the mountain belts of the Mediterranean; the Betics, Pyrenees, Alps, Apennines, Carpathians, Caucasus, Pontides, Hellenic Arc, Taurus, Zagros and Atlas ranges.

The present Mediterranean Sea is considered a successor to an earlier major ocean, the Tethys, which occupied a basin between the Eurasian and African plates from Triassic times (c. 200 Mya) through to Cretaceous and early Cenozoic times (c. 65 Mya). The evolution of the Tethys Basin has been the subject of much debate and therefore the following outline is simplified (Smith and Woodcock, 1982). In Triassic times land that was to become a part of Europe lay to the north while that of Africa lay to the south. Within the Tethys ocean, carbonate and silicate skeletal remains of marine organisms accumulated to form limestones and also the oil and gas reserves of the Middle East. Throughout the existence of Tethys crustal movements led to a changing configuration of plate boundaries and extent of the ocean. Subduction of denser oceanic crust beneath continental crust destroys original plate margins and this means that there is little relationship between the past and present coastlines, but subduction is also characterised by intense volcanic activity, for example along the present Calabrian arc of southern Italy. Where continental plates collide there is no subduction and their edges crumple.

Through the Mesozoic and early Tertiary limestone continued to be deposited, but in addition continued subduction processes and emergence of land above sea level occurred. Sub-aerial erosion of newly uplifted material resulted in coarser detrital sediments, known as flysch (beds of sandstones, shales, marls and clays), accumulating in offshore zones. At intermediate depths and distances from the shore, there was interbedding of limestone and flysch. In general flysch overlies limestone, having been deposited in basins in the limestone. With compression, the basins were uplifted more rapidly than the surrounding rocks, but as flysch is softer than limestone and is eroded more readily, it is the limestone which today forms the highest topography in many mountain regions, such as Epirus in Greece.

By 50 million years ago, the total area of the Tethys ocean was considerably reduced. Uplift of the interbedded limestone and flysch formed the mountain chains of the Alpine orogeny. In the Alps, Pyrenees and Betic ranges, the leading edges on both sides of the plate boundaries are formed of continental crust. Elsewhere, for example along the compressional belts of the coast of Cadiz, the Tellian Atlas and the Dinaric Alps, the oceanic crust of the African plate is being subducted beneath the continental crust of the European plate (Macklin et al., 1995). As well as compression, crustal movements are associated with extension, which leads to the formation of basins and small areas of oceanic crust. The rotation of Corsica and Sardinia away from France created the Ligurian basin, while the rotation of the Balearics created the Gulf of Valencia, and rifting between Calabria and Sardinia formed the Tyrrhenian Sea. In some of these basins very thick deposits have accumulated during periods of rapid subsidence, for example 10–15 km of deposit in the Tyrrhenian Sea (Ruffell, 1997).

3.1.2 Neotectonics

The study of plate movements in the geologically recent past (since the beginning of the Late Cainozoic, approximately the last 20 million years) is known as neotectonics, and neotectonic activity is regarded as instrumental in the formation of present-day topography (Vita-Finzi, 1986). Crustal movements are important controls on the landscape through fluvial systems and coastline changes, and tectonic activity continues today. Tectonic activity is not equal in intensity or occurrence around the Mediterranean. Areas with the greatest concentration of surface earthquakes are along the Italian peninsula and in the Greek islands (see Figure 3.2). Deep earthquakes are also important in these regions, and their distribution allowed Udias (1985) to divide the Mediterranean Basin into two distinct sub-regions, the western and eastern basins, separated by Italy and Sicily. Seismic activity is concentrated in the eastern Mediterranean, especially along the Hellenic Arc stretching from the west coast of Greece to southern Turkey. Here, oceanic crust of the African plate is being subducted beneath the Eurasian plate. In the western basin, there are fewer earthquakes, especially deep ones. The Mediterranean Basin is still closing, with collision zones between Africa and Arabia from Cyprus to the Zagros Mountains of western Iran, between Africa and Greece/Turkey along the Hellenic Arc, and between Africa and the Iberian Peninisula.

3.1.3 Volcanic activity

Volcanic activity in the Mediterranean is focused in two areas, along the Hellenic and Calabrian arcs. Volcanoes associated with the former include those of Melos and Santorini. Here the Ionian microplate is being subducted beneath the Aegean microplate. Santorini used to be regarded as responsible for the end of Minoan civilisation on Crete. The eruption is now dated, using several lines of evidence, to about 1630 BC, earlier than the demise of the Minoans, which is conventionally dated to around 1450 BC (Hardy and Renfrew, 1990). Volcanic activity extends northeast to the Turkish borders of the Aegean with active Neogene and Quaternary volcanoes on the island of Lesbos and in the neighbouring Kozak mountains of Anatolia (Kuzucuoglu, 1995). Volcanoes of the Calabrian arc include Vesuvius (associated with the destruction of Roman Pompei), Stromboli and the eponymous Vulcano itself in the Eolian Islands, and Etna on Sicily. Here Tyrrhenian crust is thrusting into the Adria-Apennine crust.

3.2 Structural framework of the Mediterranean region _____

3.2.1 Southern Europe

The Mediterranean region is dominated by the Alpine fold mountain system, the result of collision between the African and Eurasian plates described above

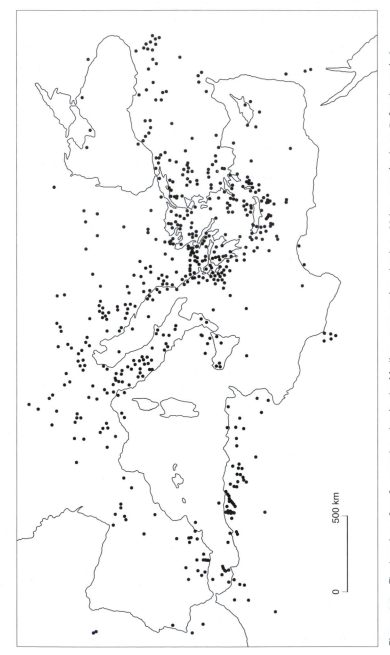

Figure 3.2 The location of surface earthquakes in the Mediterranean region, depth ≤ 60 km, magnitude ≥ 5, for the period 1910–1970. Adapted from Udias (1985).

(see Figure 3.1). Folding at this time also affected pre-existing rocks such as the Hercynian massifs of the Balkans, the Iberian Peninsula, Corsica and Sardinia. These massifs form plateaus and uplands of varied relief and complex geology (Embleton, 1984). The main phase of the Hercynian orogeny was late Carboniferous to early Permian. Today Hercynian massifs have been much eroded to create extensive surfaces of low relief. However, in some areas igneous intrusions have been important, for example Monts Dore (1886 m) in the Massif Central of France, and the mountain ranges of Corsica (reaching 2710 m). Volcanic activity was especially marked in Tertiary and Quaternary times, continuing in places through to the present.

The southern boundary of most of the Hercynian belt subsided into the Tethys geosyncline and became an extensive area of marine shelf on which limestones were deposited and later uplifted (Ruffell, 1997). In some areas, such as the Mesozoic limestone of the Dinaric Alps, karst scenery is found (see Section 3.3). However, limestones are not the only rock types; they are found in close association with highly erodible flysch and molasse sediments, some deposited during uplift and some subsequently. In deeper waters shales were deposited, which are today important sources of oil and gas. To the north of the European Mediterranean are lowlands and scarplands largely of Mesozoic and Tertiary age, some of which extend into the Mediterranean, for example those underlying the Po Valley in northern Italy.

3.2.2 North Africa

North Africa is underlain by the Precambrian African plate and, for the most part, is a desert environment. The exception is the Mediterranean region of the Maghreb, extending from Morocco to Tunisia. This is an area of Alpine fold mountains with a narrow Mediterranean and Atlantic coastal lowland (Allen, 1996). The Atlas mountains, trending WSW–ENE, dominate and can be divided into the Rif Atlas, High Atlas, Middle Atlas and Anti-Atlas mountains. The last ones separate the Mediterranean region from the Saharan region to the south and mark the edge of the African shield. At their eastern extreme in Tunisia the Atlas ranges are known as the Northern, High and Low Tell mountains. The highest elevation in the Atlas mountains is Mount Toubkal (4156 m) in Morocco. Cutting through the Atlas range are a number of major rivers, such as the Sous and the Oum er Rbia in Morocco which flow into the Atlantic, the Moulouya in Morocco which flows to the Mediterranean and the Medjerda flowing to the Mediterranean through Algeria and Tunisia. Many streams flowing north into the Mediterranean Sea have cut deeply dissected valleys.

3.2.3 The Eastern Mediterranean

Anatolia, in the northern part of the eastern Mediterranean, is again dominated by mountains of Alpine age – the Pontic and Taurus mountains. This chain continues west into Cyprus where it forms the Kyrenia and Troodos mountains. Extending south from Anatolia to the Red Sea, for about 1000 km, is the Rift Valley area (Inbar, 1998). This region includes the saline Dead Sea, at about 400 m below sea level, and the freshwater Lake Kinneret (or Galilee). Three perennial rivers also flow through the Rift Valley area – the Orontes, Litani and Jordan.

The Rift Valley separates the central mountain belt of the Lebanon, Galilee, Samaria and Judea mountains on its western side from the Anti-Lebanon mountains, Golan Heights and Jordanian mountains to the east. Elevations on both sides of the Rift Valley can reach over 2000 m, but in the central mountains the land is generally between 500 m and 1000 m above sea level (Inbar, 1998). Bordering the Mediterranean Sea is a narrow coastal lowland.

3.3 Topography and geology

The relationship between geological evolution and landforms is rarely straight-forward, especially at the regional scale. The complex folding and faulting associated with tectonism, particularly those of the Alpine orogeny, have given rise to mountainous regions and to many fault-bounded blocks and depressions, producing a basin and range topography. Long-term subsidence of inter-montane basins has led to the accumulation of thick interbedded river, lake and marsh sediments (Plate 3.2). Examples include Ioannina (Frogley *et al.*, 1999) and the Kopais depressions in Greece (Allen, 1990; Tzedakis, 1999) and Padul in Spain (Pons and Reille, 1988). Tectonic activity is a long-term control on drainage networks, with episodes of river incision and sedimentation.

Plate 3.2   The plain of Stymfalos, a fault-bounded inter-montane sedimentary basin in the Peloponnese, Greece.

Superimposed on these are climatic- and human-induced periods of landscape change – erosion and aggradation. In addition, the variety of rock types is very great and this is reflected in the landscape through differential weathering and rates of erosion. Furthermore, within a broad category such as limestone rocks there is a range of forms and, therefore, of landforms. As a result there is a wide range of Mediterranean landscapes.

3.3.1 Limestone scenery

Limestones are often thought of as the characteristic Mediterranean rock-type, associated with red, or terra rossa, soils and typical vegetation communities. As an example, the limestone region of the Algarve in Portugal is known as the *barrocal*. This has a flora quite distinct from the adjacent *litoral* (the southern, coastal non-limestone Algarve) and the inland mountainous zone or *serra*. It is in the *barrocal* that most of the vestiges of traditional Algarve agriculture are still found. In the *litoral* tourism has taken over and in the *serra* forest cover predominates (Mabberley and Placito, 1993).

Calcium carbonate limestones are the most commonly occurring limestones but in parts of the Mediterranean dolomites, or magnesium carbonates, are important (Gunn, 1986). Primary dolomite is rare; dolomitisation usually occurs as a result of diagenesis, with magnesium introduced into the calcite molecule. The relative purity or impurity of the dolomite-dominated and calcite-dominated limestones also determines their resistance to erosion, and therefore soil formation and associated landforms. Terra rossa soils (see Chapter 4) are those most readily associated with limestones but in many areas soil cover has been removed and limestone surfaces are bare. Some 2.1 million hectares of karst in the former Yugoslavia are barren, amounting to 38% of the total limestone outcrop (Woodward, 1995). The issue of soil erosion is considered in more detail in Chapter 4, so the point to emphasise here is the very slow rate of soil development on such hard rocks, with further erosion limited by lack of material to erode (Rendell, 1997).

3.3.1a Mediterranean karst landforms

Karst landforms with their characteristic solution features – sinking streams, subsurface drainage, enclosed and collapsed depressions (poljes, dolines), scarps, springs, caves and karst fields – are named after the germanic name of an area in Slovenia on the borders of the former Yugoslavia and Italy. Gunn (1986) reviews the definitions of karst and finds that the prerequisite is a rock with a high degree of solubility in natural waters. As such, karst scenery is not confined to carbonates but is also found in silicates and evaporites, such as halites, anhydrites and gypsum. However, it is generally carbonates that are dominant in karst landscapes. By definition, rocks with more than 50% by weight of carbonate minerals are limestones, and karst features tend to be found in areas of calcareous, rather than dolomitic limestones. Rocks containing dolomite are generally much less soluble than high-calcium limestones and, in addition, insoluble impurities may inhibit karstification processes. 'Ideal' karst rocks are therefore massive, well-jointed, pure, high-calcium limestones (Gunn, 1986). These are found *par excellence* in the Dinaric mountains of the former Yugoslavia, where Mesozoic calcareous rocks reach thicknesses of more than 1000 m (Bilandzija *et al.*, 1998). One reason for their extensive development is that rates of removal of limestone in solution are proportional to increasing runoff, and that rates are greater where soil and vegetation cover the rocks. This is because the amount of carbon dioxide is high in soil and vegetation mantles, and water passing through these becomes enriched in carbon dioxide and is then capable of dissolving much more limestone compared with water flowing over bare rock (Goudie, 1995). In the Mediterranean region karst landforms are best developed in regions of high rainfall such as the former Yugoslavia.

Limestone landscapes are not just found within the former Yugoslavia but are widespread throughout the Mediterranean. Caves developed within limestone ranges provide evidence of their use as rock shelters dating well back into prehistory, such as those in the Dordogne area of southern France. Cave sediments contain the remains of their bird, animal and human populations and so are a prime source of prehistoric information. An excellent example of this is provided by the rock shelter at Klithi in the Pindus mountains of Epirus, Greece. Klithi is one of at least 26 such rock shelters in the limestone region of the Voidomatis gorge (Bailey, 1997). Deposits within the shelter are littered with flint artefacts and bone assemblages which have provided invaluable information about the human occupancy of the site. At the height of the last glaciation (some 20 000 years ago in this area) permanent snow covered the mountain slopes of the Pindus. This began to retreat at the same time as the first occupants arrived at Klithi, although people had been in the region for at least the last 100 000 years. These occupants must have witnessed the change from the cold arid late glacial conditions to those more akin to the present – warmer and wetter. They lived there until about 10 000 years BP, although the cave continued to be used as a winter goat shelter right up until recent archaeological investigations of the site.

Limestone catchments are also aquifers with substantial reserves of groundwater and so are major sources for irrigation water – a necessity in a semiarid climate. However, water quality is at present declining and is the subject of much concern. Quarrying of limestone for cement is a further cause for concern in terms of landscape conservation. In some areas there is a long history of quarrying, such as marble from the famous Carrara quarries in Tuscany, Italy. Today extraction rates are nearly equal to the maximum reached during Roman times and in the past forty years some 55 million cubic metres have been quarried. In addition, a volume so far of more than 100 million cubic metres of spoil has been created (figures quoted by Conacher and Sala, 1998).

3.3.2 Other rock types

Although limestones are a major component of bedrock types in the Mediterranean region, they are not the only rock type and they contribute relatively

Plate 3.3   Badlands scenery near Thebes, Boeotia, Greece.

little to the suspended sediment load of rivers. Other common rock types are marls, shales, flysch and volcanics, some of which are more easily eroded than limestone. A review of erosion rates around the Mediterranean Basin shows that some of the highest rates occur on marl slopes, especially in devegetated areas (Conacher and Sala, 1998). In the Iberian Peninsula erosion rates were reported as 266 and 454 t/ha/yr from marl and clay slopes, respectively, compared with losses of 1.1 and 7.3 t/ha/yr from burnt phyllite and granite surfaces, respectively. In areas like the Ebro basin in Spain, marls are associated with rilling, gullying and badlands topography. Field experiments show that runoff rates and sediment losses are consistently higher on bare marl soils compared with shale-sandstones or conglomerates. On soils formed on marls sediment losses averaged 491 g/m^2 compared with 21 g/m^2 and 176 g/m^2 for the other two, respectively. Soils formed on marls may be deep but due to their lower than average ability to hold moisture, runoff rates and soil loss are higher than on sandstones or conglomerates with similar gradients and slope management practices. In dry years, shale soils cannot support a vegetative cover and so are especially susceptible to erosion. The presence of surface rock fragments then becomes important in reducing soil water evaporative losses. However, on many Greek marls coarse rock fragments are absent, which increases the risk of soil degradation (Kosmos and Danalatos, 1998).

Areas of extensive outcrops of shales, clays and flysch deposits may form badlands scenery (Plate 3.3). The term badlands applies to intensely dissected landscapes where vegetation is sparse or absent. These regions are generally useless for agriculture. Badlands are found in many Mediterranean countries. In Spain, rates of erosion are high in the badlands areas of the Ebro Valley and Valencia (Conacher and Sala, 1998). Measurements with erosion pins in the Ebro have shown an average ground lowering of 3–9 mm per year in an infilled valley and 8–17 mm per year on a slope. In some areas of Valencia erosion may be as high as 50 mm per year and rates are spatially variable. Erosion appears to be directly related to rainfall during the periods of measurement – some 87–92% of the total sediment production in the infilled Ebro erosion plots was generated by runoff processes. Gullies in these badlands seem to develop from collapse of subsurface channels, or pipes. In semiarid environments pipes may be several metres in diameter and carry water and sediment. They are most commonly found in steep slopes and along gully sides. Their presence is often regarded as an indication of soil degradation and accelerated erosion.

In some areas pipe networks have developed over the last few centuries, as reported for the Almanzora basin in Spain (Conacher and Sala, 1998) but elsewhere they may be even older. In southeastern Spain, archaeological materials some 4000 years old have survived *in situ* on badlands slopes (Wise *et al.*, 1982). The region of the Guadix basin has spectacular badlands scenery with high drainage densities and steep valley side slopes which are only sparsely vegetated. Given the semiarid climate, the expectation is of high rates of erosion. Yet estimates of erosion losses are only moderate and the *in situ* archaeological deposits suggest greater slope stability and less rapid erosion than an initial visual impression would give. The long-term denudation history of the area suggests continuing incision along the drainage lines since Pliocene times in response to tectonic activity, but very slow development and erosion of the interfluve areas on which the archaeological material is found. If such 'old' badlands are found elsewhere around the Mediterranean then it indicates a need for a cautious approach in assuming that all badlands are the result of recent accelerated erosion.

Where there are outcrops of bare flysch erosion rates may also be high and associated with landsliding. Steep flysch slopes are unstable and do not support a continuous vegetation cover. This makes them susceptible to rapid runoff and erosion after high-intensity storms, often initiating gullying and a badlands-type scenery, for example in the Epirus region of Greece (Bailey, 1997). Here in the Voidomatis catchment modern alluvial sediments are primarily flysch in origin from a mixed limestone and flysh catchment (Woodward *et al.*, 1995).

Attempts to trace source areas of river sediments involve analysis of their physical, chemical and mineral magnetic properties and comparing these with soil samples from differing parts of river catchments. Mineral magnetic properties have the potential to differentiate between topsoil and subsoil sources of sediment (Thompson and Oldfield, 1986). In the Algarve, southern Portugal, James and Chester (1995) were able to demonstrate that 'younger fill' sediments (see Section 3.4.1) could be grouped according to catchment geology; it is possible to distinguish between sediments originating from limestone areas and those from 'metasediments' of the *serra* upland comprising flysch facies, shales, phyllites and greywackes. These are the chief source of river sediments. The magnetic susceptibility of the sediments suggests a topsoil source consistent with overland flow and superficial erosion of the upland soils as a result of land disturbance and vegetation clearance during the past three thousand years. Widespread removal of *Quercus* (oak) forest is associated with the Moorish occupation of the Algarve from the ninth to twelfth centuries AD. Soil erosion at this time led to the silting of the Arade river, which had previously been navigable to the sea and a route for the export of timber. Silting of the river is believed to have been partly responsible for the decline of the town of Silves, a former capital of the kingdom of the Algarve.

Where limestone is present in geologically-mixed catchments it can exert an influence on channel discharge and sediment transport in the form of an on/ off switch. The discharge of streams flowing over non-limestone bedrock onto limestone decreases as the water disappears below ground. In consequence sediment transport reduces or even ceases. This reduction in discharge and its effect on channel geometry has been demonstrated for some catchments, again in the Algarve (Bennett, 1997). Both channel width and depth decline by some 20% as rivers flow from *serra* rocks onto limestone. Any *serra* material being transported is deposited – erosion of further material and its transportation only occurs again as rivers flow off limestone and through localised deposits of Pleistocene sands (Allen, 1999).

3.4 Drainage development and landscape change

Modifications to vegetation cover within river catchments and changes in water balance alter runoff rates, soil erosion and suspended sediment yields. Relatively few Mediterranean rivers have long-established records of sediment yield but information for some selected catchments is shown in Figure 3.3a (Macklin *et al.*, 1995). Highest yields are found in catchments that are prone to high-intensity rainfall events and have steep relief, unconsolidated sediments and friable soils. The very high figures for the former Yugoslavia and Albania are associated with severe sheet and gully erosion. Once the natural vegetation is modified, suspended sediment yield tends to increase. However, sediment yield data only provide a partial picture of the amount of erosion taking place; much of the material eroded within a catchment never reaches a river and is not transported to the outlet of the drainage basin. It accumulates as slope and alluvial deposits, as described for some Algarve rivers in the preceding section. In addition, the magnitude of sediment yield reflects other factors, such as tectonic activity, topography and lithology. Nevertheless, several studies have shown the influence of mediterranean-type climates on sediment yield (Dedkov and Moszherin, 1992); for mountain regions, sediment yield in the Mediterranean zone is second only to that in glacial zones (see Figure 3.3b). Once the anthropogenic component of sediment yield is considered the vulnerability of mediterranean zones to disturbance and erosion becomes even more apparent. Removal of vegetation reduces evapotranspiration and increases surface runoff during storms. This may extend the drainage network and lead to piping, rilling, gullying and accelerated erosion. At the same time, soil compaction reduces soil depth and soil water storage and infiltration capacity. As a result groundwater recharge is less and there are lower river flows during the dry season.

from post-Roman times. From various localities it has been described as buff-coloured or greyish-brown silty sand. Vita-Finzi's dating relied upon the stratigraphic position of the fill with respect to archaeological finds, such as pottery, embedded within the fill. Despite its clear associations with archaeological remains, Vita-Finzi believed that the fill was the result of a change in the power of streams to export sediment from their basins, and that this reflected a change in climate. Part of his reason for rejecting an anthropogenic origin for the fill lay in the argument that Mediterranean landscapes had long been the province of human activity, not least in Roman times, yet at the time of Vita-Finzi's writing there was little evidence for valley aggradation in Roman times and very few radiocarbon dates were available for material recognised as 'younger fill'.

The debate over a climatic as opposed to an anthropogenic origin for the younger fill was vigorous through the 1970s and continued into the 1980s. However, it is now commonly accepted that many different Holocene fills exist, dating from different times at different places, and ascribing these to climatic change is too simplistic (Wagstaff, 1981). Nevertheless, Vita-Finzi's work stands as a 'classic' in physical geography (Grove, 1997) and the terminology of 'older' and 'younger' fills is still used, nor has the role of climate in Holocene fluvial systems been completely dismissed.

Undoubtedly, anthropogenic activity has had a marked effect on Mediterranean landscapes. Rivers respond sensitively to changes in sediment delivery rates to valley floors, and modification – especially reduction – of vegetation cover within drainage basins tends to increase runoff, soil erosion and sediment delivery to streams and rivers, and hence increase suspended sediment yield. This has long been recognised by archaeologists and geomorphologists, who regard river sediments as important sources of evidence in trying to unravel the interrelationships between prehistoric human activity and environmental change. An excellent example of this is a large programme of geomorphological and archaeological work in the Bifferno Valley, in the Molise region of Italy (Barker and Hunt, 1995). Such an integrated approach is important because the nature of former fluvial systems is of interest to archaeologists trying to understand resources (including land) available to former settlements and civilisations, while for geomorphologists, archaeological evidence can provide dating control from material buried within or found *in situ* upon alluvial sequences. In addition, valuable information about how land was managed (for example terracing and flood control) can help geomorphologists to understand landscape changes. Caution is needed, however, with any such approach to ensure that there is no circularity of argument – disentangling competing hypotheses of landscape change is vital to reconstructing past changes and to understanding the present dynamic landscape.

In the Bifferno study early valley alluviation occurred in the Pleistocene in response to both tectonic and climate change, while during the Holocene episodes of alluviation coincided with intensification of settlement and land use (Barker and Hunt, 1995). In particular, archaeological surveys point to expansion of rural settlement and intensification of land use between the third century BC and the first and second centuries AD. This correlates with considerable aggradation at the time. Sediment characteristics suggest soil erosion caused by extension of cultivation. Similar settlement expansion/aggradation phases are suggested for the tenth to thirteenth centuries AD and for the 'early post-medieval' period of the sixteenth to eighteenth centuries. Human land-use practices are regarded in Bifferno as having been a major agent of landscape change. Indeed, the combination of anthropogenic disturbances and climatic fluctuations may increase the susceptibility of some river systems than either factor acting alone.

3.4.2 Landscape change – the role of tectonic activity

While the influences of climate and anthropogenic activity have long been recognised, the appreciation of the role of tectonic activity in determining drainage development and landscape change, especially in fluvial systems, is more recent (Vita-Finzi, 1986). Uplift leads to a change in base level, and this leads to drainage network modifications (incision and river capture). Folding, faulting and tilting have significantly disrupted drainage networks, including drainage reversal, stream capture, dissection or ponding (Macklin *et al.*, 1995). In tectonically active areas, such as those bordering the Gulf of Corinth in southern Greece, river catchments have developed in response to slopes generated by normal faulting (Collier *et al.*, 1995). Some of the large normal faults are 10–20 km in length, and in any single seismic event there may be displacement of between one and a few metres. In 1981 three earthquakes with magnitudes greater than six shook the eastern Gulf of Corinth and during one of these part of the coast sank by 1.0 m (Vita-Finzi and King, 1985). Such a scale of movement affects archaeological and geomorphological records of landscape change. Along the Gulf displacement varies between 0.3 mm/yr at the eastern end and 1.6 mm/yr at the western end. Antecedent drainage basins have responded to uplift by elongating and downcutting into uplifted areas downstream, and knickpoints have gradually 'migrated' upstream. Other rivers have maintained their courses where they cut through easily erodible Neogene deposits. This latter point emphasises the need for caution in explaining landscape activity. For example, local differences in lithology can override regional histories of tectonic activity. Nevertheless tectonic modifications may be significant in landscape and drainage basin development.

Volcanic activity may involve the eruption of lava and ash. Studies of the Alcantara and Simeto rivers which drain Mount Etna have shed light on the relationships between volcanic and fluvial events (Chester and Duncan, 1982). Prior to the onset of Etna's activity in Middle Pleistocene times, the areas now occupied by the lower reaches of the rivers were near-shore marine environments. Several episodes of activity followed, as indicated by different lava sequences, with periods of volcanically induced sedimentation and erosion. Aggradation occurred during periods when eruptions produced sediment-choked streams. These were separated by periods of non-deposition and incision as streams adjusted to a diminished sediment load.

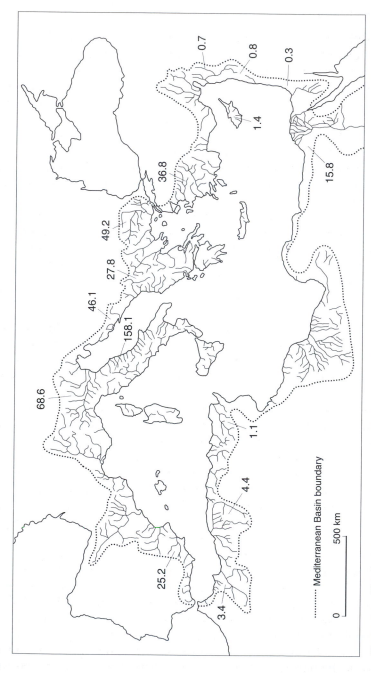

Figure 3.4 The river drainage network of the Mediterranean Basin. Values given on land indicate river runoff and agricultural discharges. Adapted from Maklin *et al.* (1995) and Cruzado (1985).

3.5 Hydrology

3.5.1 Flow regimes

A marked difference exists between river flows on the northern shore of the Mediterranean and those of the eastern and southern coastlines (Figure 3.4). In general the European coastline is relatively well watered and river flow is perennial. Figures for annual runoff are much greater than those further south and east. Some rivers, like the Rhône, originate outside the Mediterranean-climate zone and so their hydrology is, to a large extent, determined by climatic events outside the region. Other rivers, like the Ebro and Po, have their head-waters in mountain regions and are fed by melting snows. In contrast many of the drainage networks of North Africa and the eastern Mediterranean are poorly developed, and unless water is derived from outside the Mediterranean-climatic zone, as in the River Nile, rivers are characteristically ephemeral. Fewer river systems are found in the more arid southern part of the Basin.

Given the scale of the Mediterranean region and the problems in monitoring discharge and the complexity of the relevant equations, attempts to calculate continental runoff are fraught with difficulties. Nevertheless some estimates have been made (Cruzado, 1985). Their utility lies in the need to determine the water chemistry and threats to changing water quality of the Sea itself. The flow of fresh water from river runoff and other sources into the Mediterranean Sea is shown in Table 3.1 and Figure 3.4. These demonstrate the imbalance between the northern, and southern and eastern shores. The region which receives the largest volume of freshwater is the Adriatic, followed by the north-western Mediterranean, the Aegean Sea, the Tyrrhenian Sea and the Ionian Sea.

The annual river regimes largely reflect the nature of the Mediterranean climate and its variation across the Basin, with peak flows in late autumn, winter and early spring, and minimum flows in summer (Macklin *et al.*, 1995). Where rivers rise in mountainous regions, as in the Balkans, Anatolia and the Iberian Peninsula, there may be strong early spring peak flows as snows melt. Flooding often occurs in summer as a result of thunderstorm activity, frequently enhanced by relief rain, but its occurrence is generally localised. However, flooding may also be common, and sometimes catastrophic, in the autumn (see for example Section 2.5).

In response to climatic changes during the cold stages of the Quaternary, river flow would have been very different from that of today and, indeed, it is predicted to change in response to contemporary global warming. Indications of changing hydrological regimes come from studies and dating of fluvial sediments. For the Quaternary cold stages, river regimes were probably more seasonal than at present and exhibited greater spatial and temporal variability (Macklin *et al.*, 1995). This was especially so of rivers flowing from mountainous regions in the northern Mediterranean, where there was increased snow cover. Some northern catchments would have been glaciated (see Section 3.6) and in these winter would have been the season of minimum flow as snow and

Table 3.1 Flow of fresh water from river runoff and other sources into the Mediterranean Sea

Region	Approx. area (10³ km²)	Total (m³s⁻¹)	Specific discharge (mm)	Percentage contribution from				Population (× 10³)
				Rivers	Domestic	Industrial	Agriculture	
Adriatic	133	5 195	1228	66.2	0.2	0.7	32.9	3 625
Northwestern	287	3 031	331	84.5	0.8	2.5	12.2	8 870
Aegean	143	1 463	321	58.1	0.4	1.5	40.0	4 584
Tyrrhenian	231	1 225	167	38.0	1.1	1.8	59.1	8 131
Ionian	151	1 103	230	21.3	0.3	0.9	77.5	1 883
North Levantine	177	778	139	84.5	0.1	0.1	15.3	1 320
South Levantine	500	550	35	89.3	1.1	0.8	8.8	5 388
Southwestern	264	290	35	39.0	1.2	2.8	57.0	4 439
Alboran Sea	68	197	91	61.0	1.9	2.3	34.8	2 690
Central	583	105	6	0.0	2.8	3.9	93.3	2 904
Total western Basin	850	4 725	173	68.6	1.0	2.3	28.0	24 130
Total eastern Basin	1687	9 194	170	61.7	0.3	0.8	37.1	19 704
Total Mediterranean	2537	13 919	173	65.2	0.5	1.4	32.9	43 834

Source: Cruzado (1985; from 1977 United Nations Environment Programme data).

ice accumulated and then thawed to give a late spring/early summer peak discharge. Evidence of increased sedimentation in the cold stages of the last glacial period comes from the Voidomatis catchment (Macklin *et al.*, 1997). Sedimentary units dated to between 24 300–26 000 and 28 200 BP are suggested to correlate with a severe cold period revealed by the oxygen isotope record of Greenland ice cores. This is believed to be the result of massive injections of cold water with the release of icebergs into the North Atlantic. This is known as the Heinrich 3 event (Bond *et al.*, 1992). The influence of cold surface North Atlantic waters and cold air over Greenland spread into Europe and the Mediterranean. Frost weathering and break-up of rocks in the Pindus mountains, among other regions, would have increased sediment delivery rates.

In the eastern Mediterranean, the late Holocene has been more arid than the period from 14 000 to 5000 years BP and, as a result, runoff and denudation rates have been lower. Alluvial fans and fan deltas around Greece, which were formed during the early Holocene, are now typically undergoing incision as sediment yields are reduced compared with earlier periods. This is evident in catchments that have not been modified by anthropogenic effects such as deforestation and increased grazing (Collier *et al.*, 1995).

Long-term climatically induced changes in sediment delivery rates are also evident in marine deposits. In the eastern Mediterranean thick layers of black mud rich in organic matter, known as sapropels, have been identified (Rossignol-Strick *et al.*, 1982). These are ascribed to periods of high Nile discharges of freshwater, which form layers of low-salinity surface water. The bottom water in this part of the Sea is highly saline and the steep salinity gradient resulted in stratification of the water column. The bottom waters were depleted of oxygen and in these stagnant conditions organic-rich muds accumulated. The periods of high discharge are believed to be a consequence of increased monsoon precipitation and increased runoff from the Ethiopian highlands, the source region for the Blue Nile and its tributary, the Atbara. The two most recent sapropels have been dated to between 11 800 and 10 400 years BP and to between 9000 and 8000 years BP, times of intensified and northward displacement of the African monsoon during the late glacial to Holocene transition. The interruption to their deposition between 10 400 and 9000 BP is believed to result from a drier equatorial phase during the cold period, known as the Younger Dryas (Rossignol-Strick *et al.*, 1982). At this time large volumes of cold water from melting ice sheets in North America and Europe flowed into the North Atlantic, temporarily shutting down the Atlantic 'conveyor-belt' of ocean currents (Broecker, 1994).

As water plays a vital role in all societal activities and landscape characteristics, any future changes in the hydrological cycles and availability of water resources as a result of climate change are worth examining. However, there are considerable uncertainties involved in estimating the effects of climate change on river runoff and in the translation of these effects into actual impacts in terms of water supply, power generation, navigation and the aquatic environment (Arnell, 1999). Some of the uncertainties lie in the parameters of

The major coastal lowlands have provided extensive areas for settlement and agriculture. The initiation of the Predynastic occupation of the Nile delta (7000 to 5000 years ago) occurred relatively soon after a deceleration in rising sea levels about 7500 years ago (Stanley and Warne, 1993a, b). This led to the accumulation of silt and a widespread and fertile plain which buried early Holocene sandy deposits. The new seasonally flooded delta developed an increasing vegetative cover and provided a setting conducive to the establishment of agriculture, which spread from the eastern Sahara. Farming and herding replaced hunting and gathering. As well as a halt in rising sea levels, changing environmental factors aided the establishment of the Predynastic cultures. It is believed that severe drought in the eastern Sahara encouraged farmers to move from there into the Nile region. The changing pattern of sedimentation, with a transition from sandy to overlying silty deposits, as a result of a deceleration in sea-level rise, is matched by similarly dated lithological changes in the Ebro and Rhône deltas (Stanley and Warne, 1993a).

Before construction of the Low Aswan Dam in 1902 and the High Aswan Dam in 1964, the Nile transported more than 120 million tonnes of sediment annually to the sea, with an average discharge of about $84 \times 10^9 \text{ m}^3$ per year. In addition, more than 9 million tonnes of suspended sediment were deposited on its flood plain each year, equivalent to an annual layer of silt about 1 mm thick (Stanley and Warne, 1993b). Today the existence of the Aswan dam has greatly reduced these figures. Flow downstream of the dam has become constant rather than seasonal, which allows irrigation through the year. Some Nile water is also diverted for agriculture along the valley and in the delta region. Water is also taken for the growing population of Cairo. Some 98% of sediment is trapped by Lake Nasser, behind the High Dam. The only new sediments reaching the delta arrive by longshore drift from the Arabs Gulf and Alexandria shelf. As a result the delta is diminishing in area. There is accelerated erosion along the delta coastline and marine incursion onto the low-lying areas of the delta. Soils in the delta region are becoming more saline as annual flooding no longer flushes out evaporitic salts.

Although impoundment of Nile waters has had a clear impact on Nile hydrology there is some evidence that climatic changes may also be responsible for shore migration trends in the delta (Frihy and Khafagy, 1991). Surveys of the delta show two distinct trends – prograding shorelines during the nineteenth century and regression since 1900 associated with particularly high flows in the nineteenth century and significantly reduced flow in the twentieth century. Although the Low Aswan Dam was built in 1902 it was designed to allow flood waters to pass through. Frihy and Khafagy therefore argue that reduced twentieth-century flows are related to climate change as well as to dam construction. Flow data from upstream of the Aswan dams has been correlated with Nile discharge and water levels in lakes Naivasha and Turkana in Kenya, and Lake Chad. This suggests that years of high Nile floods are the result of tropical monsoon rains. In contrast, reduced discharge in the early twentieth century is probably the result of aridity and the regulation of water flow by dams.

The range of different deltaic ecosystems around the Mediterranean Sea – lagoons, saltmarshes, freshwater marshes, beach–dune complexes – provides important habitat for animal and plant life. Today much of this natural and semi-natural habitat is threatened by reclamation, development and pollution (Hollis, 1992). For example, waste water from Nile irrigation projects and industry in the Cairo region is generally polluted. It is discharged into coastal lagoons where it damages the natural habitats for fish and migrating birds (Stanley and Warne, 1993b). However, coastal wetland environments are generally transient. Some changes are seasonal, annual or decadal in scale. Others can be observed as sea levels change over the range of glacial and interglacial cycles and within interglacial periods such as the present Holocene.

3.7.1 Sea-level changes

The advance and retreat of ice sheets during the late Pleistocene was accompanied by major eustatic sea-level changes as water was removed and returned to the oceans and by isostatic changes associated with redistributions in the surface loads of ice and water. Reconstructing sea-level changes is complex, relying on multidisciplinary evidence and detailed chronological control. At the Last Glacial Maximum, some 18 000 years ago, there is much bathymetric evidence that suggests a sea stand of between −90 and −130 m (Pirazzoli, 1998), though many different estimates exist. As the ice melted sea level rose from about 15 000 years BP until mid-Holocene times. Since then many researchers believe that sea level has been relatively stable, though some argue for small oscillations in level while others suggest a continued rise (see Lambeck, 1995 for further discussion on this). To determine sea-level changes researchers use records of vertical land/sea movements from tide gauges and archaeological, geological and geomorphological evidence. The challenge to produce a unifying global curve of sea-level change is futile but there have been attempts to analyse the differing contributions of eustatic, isostatic and tectonic components at the regional scale, such as in the eastern Mediterranean, and to assess the influence of sea-level changes on people and their environments.

Theoretically there is good scope for the use of tide gauge data to reconstruct sea-level change in the Mediterranean because the tidal range is small and there are scattered bays in which storm effects can be neglected (Vita-Finzi, 1986). However, within the Mediterranean there are only 16 published long records (50–100 years) and 15 short records (10–20 years) (Flemming, 1992). These are supplemented by some 335 good estimates from archaeological records – the submergence of archaeological sites, or the raising of ancient harbours, above the present level of the sea. However, some coastlines are over- or under-represented with respect to available data and extrapolations of land/sea movements should not be extended over distances greater than 10–20 km. Nevertheless there are some well-preserved coastal features which can be correlated across long distances. For example, on the coastline of western Crete there is geomorphological evidence in the form of notch marks 6.8 m above the modern low-water line on the Gramvousa peninsula (Kelletat, 1991).

On steep, rocky slopes newly exposed areas are only a few metres wide, but elsewhere coastal plains up to 500 m wide have been exposed with archaeological remains dated to about 2000 BP found on the raised surfaces.

To make sense of reconstructed sea-level changes in the eastern Mediterranean, Lambeck (1995) has developed a model of eustacy and isostacy for tectonically stable areas of the world. This glacio-hydro-isostatic model is based on the fact that relative sea-level change is a function of eustatic sea level, an isostatic correction and a tectonic contribution. The eustatic component comprises the loss of water from the oceans to ice during periods of glaciation. The change in sea level is also affected by the gravitational attraction between ice and water. The gravitational attraction of newly formed ice tends to pull water towards the ice mass. Therefore close to an ice sheet, sea level rises relative to the eustatic change (which is negative) during times of increasing ice mass, but in order to preserve a balance, further away from the ice sheet sea level falls by more than the eustatic amount. Changes in the distribution of ice and water cause adjustments in the Earth's lithosphere which overlies the mantle – elastic deformation occurs in the lithosphere, while there is viscous flow in the mantle. As ice sheets melt so the surface load alters, producing an isostatic response. The formerly glaciated crust is unloaded while the meltwater added to the oceans loads the oceanic lithosphere. The degree of change and the relative importance of the two opposing adjustments depends on distance from the former ice sheet. In regions near to previously glaciated areas crustal adjustment is one of uplift and sea level appears to fall. Further from the formerly glaciated regions subsidence becomes more important and sea level apparently rises. The isostatic response extends well beyond the formerly glaciated regions and Lambeck has shown the influence of the Fennoscandian ice sheet in the eastern Mediterranean. Using the glacio-hydro-isostatic model, he has demonstrated that rising sea levels around much of the Peloponnese in Greece are a consequence of isostatic readjustment.

Sea-level changes during the Late Pleistocene and Holocene would have been of major significance for the inhabitants of the Mediterranean. Increased areas of coastal plain would have facilitated local travel along coastlines and migration routes for people and animals across longer distances. In addition, more land would have been available for habitation, hunting, gathering and grazing of animals away from the interior mountains. It has been suggested that such an increased resource base provided a stimulus for a transition towards a much larger subsistence economy than that which existed before, and this is suggested to have happened in later Palaeolithic times, some time during the Last Glacial Maximum. Accordingly, van Andel (1989) presented reconstructions of coastlines for two time-slices: the first during the Last Glacial Maximum between 24 000 and 14 000 years ago when sea levels were at least 100 m below that of today, and the second at about 9000 BP at the end of the rapid post-glacial sea-level rise, when the level stood about 35 m below the present level. These reconstructions are shown in Figures 3.5a and 3.5b. For the late glacial coasts (Figure 3.5a) Anatolia was connected to Europe

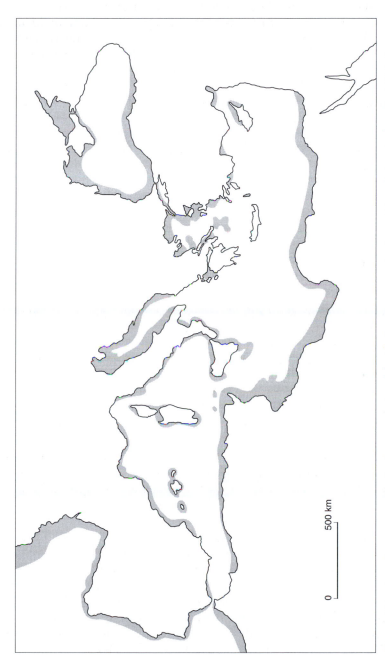

Figure 3.5a Late Quaternary coastal palaeogeography of the Mediterranean: sea level at −120 m during the Last Glacial Maximum. After van Andel (1989).

0 | 500 km

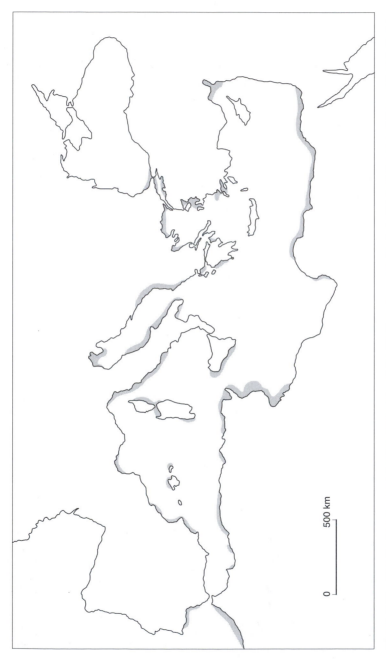

Figure 3.5b Late Quaternary coastal palaeogeography of the Mediterranean: sea level at −35 m at the end of the rapid post-glacial sea-level rise (early Holocene). After van Andel (1989).

across the Bosporus, the islands of the Cyclades equalled one island, and large expanses of coastal plain existed in the present northern Adriatic and Gulf of Gabès, and fringed most of the Iberian Peninsula, southern France, Italy and Greece. The position of the deltas of the Ebro, Rhône, Po and Nile had all shifted seawards, with a changed rate and pattern of sediment delivery to the shelf-slope depositional area (Macklin *et al.*, 1995). The Strait of Gibraltar remained open, even though only 8 km wide. Italy and Sicily were joined by a narrow land bridge, but as its depth is located at 90 m below present level, it did not survive long once the deglacial phase began. Overall, the distances by sea from one coastal location to another, for example between Sicily and Tunisia, or Crete and mainland Greece, were much reduced. By about 9000 years ago the present coastline of the Mediterranean became more or less established (Figure 3.5b). The distance between Sicily and Tunisia increased to 200 km, and that between Italy and Corsica to 60 km. The coastal plains of Tunisia and the Adriatic disappeared.

During the Holocene sea levels have not been at a constant elevation with respect to the land. This has implications for topography, archaeology and the distribution of animal and vegetation communities. An example of a detailed study of marine incursions onto a coastal plain comes from the Argive Plain and Gulf of Argos, on the Aegean Sea. Van Andel *et al.* (1990b) used borehole logs and marine seismic reflections to correlate Quaternary oceanic and terrestrial sedimentary sequences. These revealed episodes of alternating marine and fluvial deposition accompanied by soil formation at the time these deposits were accumulating. Marine deposits within the Holocene material extend up to 1.5 km inland of the present coastline, indicating the maximum extent of the post-glacial transgression. Today the coastline is regressive. Many sites of historical and archaeological interest in Greece and Turkey which were formerly on the coast are now located inland, such as Troy in Anatolia and the battle site of Thermopylae north of Athens.

In many deltas compaction of Holocene sediments, often accompanied by extraction of groundwater and oil or gases from below ground, is leading to subsidence and therefore a relative rise in sea level. This is the case in the Po delta, Italy where natural subsidence rates are of the order of 1 mm per year, though these have been exacerbated by reclamation of coastal sediments over the past 400 years (Pirazzoli, 1998). This involves pumping out water. In addition, extraction of methane from 1938 to 1964 increased the rates of sinking. These levelled out when methane extraction stopped, but through the 1970s rates of subsidence were between 5 and 20 mm per year. All the land in the delta region today is below sea level. Since the 1950s many parts of the shoreline have been retreating (Sestini, 1992). This region includes the city of Venice, initially settled in the sixth century AD because of the protection afforded by the natural lagoons and their swampy margins. The continued existence of the city depends on the ecological balance of its lagoons. Sea defences built to protect the agriculture and tourist infrastructure of the region also mean that the natural flexibility of the coastline no longer exists. Increasingly Venice is subjected to flooding, first brought to international attention

in November 1966 when 1.94 m of water inundated the city. As a result
several policy initiatives and laws have been implemented to protect Venice,
not all of which appear to be succeeding, particularly in the light of continued
sea-level rises as a result of CO_2-induced global warming (Dardis and Smith,
1997).

3.7.2 Sea-level changes and tectonic activity

A complicating factor in the study of sea-level changes is that changes in the
relationship of land and sea may be due to local tectonic activity as well as
eustatic and isostatic activity. Tectonic activity is, of course, not uniform
around the Mediterranean Basin. Therefore records and research findings on
sea-level changes must not be applied uncritically from one area to another,
either from the eastern to western coastlines or even across the eastern Mediterranean coastline, which is one of the most rapidly deforming areas on Earth.
Lambeck (1995) applied the glacio-hydro-isostatic model to reconstructed
sea-level records from the eastern Mediterranean to determine the likely impact
of tectonic activity on the basis that differences between the observed sea levels
and those predicted by the model can be attributed to tectonic factors. For the
southern Peloponnese, the role of tectonic activity in the Holocene relative
sea-level rise is discounted, partly because shorelines associated with the Last
Interglacial occur only a few metres above the present shoreline. If the Holocene
sea-level changes were tectonic rather than isostatic then the Last Interglacial
shorelines would be some 100 metres below present sea levels. However, the
minimal influence of tectonic activity in the southern Peloponnese should
not be extrapolated to other eastern Mediterranean areas. Application of the
glacio-hydro-eustatic model to other parts of the eastern basin shows that in
some places observed sea-level changes are greater than those predicted by the
model, particularly in western Crete and Rhodes, and indicative of tectonic
uplift. These localities are part of an arcuate zone of uplift extending from
Rhodes in the east to Kithera in the west (in the southern Aegean) and
incorporating Crete. Rates of uplift locally exceed 4 mm per year and uplift of
as much as 10 m is estimated to have occurred on the Gramvousa peninsula
(mentioned in Section 3.7.1) associated with a tectonic event dated to 1550
BP (Kelletat, 1991).

   Earthquakes may be associated with tsunami which inundate coastal regions
and their ecosystems. For example, the eastern section of the Azores–Gibraltar
plate boundary is a zone of active plate compression, the location of many
deep-seated earthquakes (Dawson et al., 1995; Hindson and Andrade, 1999).
The so-called Lisbon earthquake of AD 1755 was triggered by movement along
these plates. While earthquake damage and fire destroyed much of Lisbon, a
tsunami caused flooding and fatalities along the coastlines of Portugal, Spain
and much of Morocco. At Boca do Rio in the Algarve, Portugal, a distinct
sedimentary horizon occurs within saltmarsh deposits. Its sandy nature suggests that the main source of material was from the beach–foredune system,
seaward of the marsh. As the wave crashed onto the shoreline it eroded beach

sands, and then redeposited these inland (Hindson and Andrade, 1999). It is possible that the wave caused significant erosion of pre-existing sediments. It also led to a change in the ecological status of the wetland as the barrier created by tsunami sediments changed the wetland from being a saltmarsh to a fresher water marsh as river inflow was impounded (Allen, 1999). Gradually the barrier was breached, returning the marsh to its present saline state. Further east along the Algarve coast, the tsunami caused severe damage to the chain of barrier islands of the Rio Formosa (Andrade, 1992).

3.8 Further reading

Van Andel, T.H. (1994) *New Views on an Old Planet*, 2nd edition, Cambridge University Press, Cambridge.
Vita-Finzi, C. (1986) *Recent Earth Movements*, Academic Press, London.

Chapter 4

Soils

'The first sight and last sight of Greece, as the traveller arrives or departs will be of mountains, of hills, of bare rocky slopes climbing from the sea to high ridges and crests. There is nothing here of softness and obvious charm. There is beauty in plenty, but of a stern uncompromising type where weathered rock and stone are set at such an angle that soil has scarce chance to collect, and where it would seem at first sight there is nothing to offer for human settlement and cultivation.'

Huxley and Taylor (1977, p. 19)

Yet soil does form and collect in upland areas, among the rocky bones so near the surface, and people have lived in the mountains for thousands of years successfully cultivating crops. In spring many hillslopes are ablaze with flowers between trees grown on terraces to trap soil. In places soil is being and has been eroded and transported from the mountains to the plains. These level and low-lying areas have been the focus of settlement and have long been used extensively for agriculture, with recent intensification. Soil degradation has occurred and still is occurring. The same picture can be found throughout much of the Mediterranean Basin and the same questions recur. Are Mediterranean soils resilient or fragile? Are they especially prone to erosion, whether caused by natural or anthropogenic processes? In turn these questions beg others. In particular, are soils typical of the Mediterranean unique to the region or widely found elsewhere? Red or terra rossa soils are often found in the Mediterranean. What makes these soils red? This chapter aims to answer these questions by examining the factors which cause soils to vary – to what extent are soil properties in the Mediterranean region a function of rock type, climate or time, or a combination of these?

Mediterranean soils are highly varied and it is misleading to generalise about them, but some common characteristics are that they may be thin and poorly differentiated in terms of horizon development, that they may have developed slowly, that they may have poor nutrient content and differing organic carbon content, that they may be stony, red, often calcareous, and in places subject to salinisation. Such characteristics may increase the potential for their degradation. Erosion of topsoil on limestone substrates, for example, is mostly of organic matter rather than the mineral matter. It can then take many years for the organic matter to be replaced naturally. Soil erosion affects plant and animal communities, but these communities also influence the soil – its formation and susceptibility to erosion. The 'architecture' of different vegetation communities and differing land-use patterns can either hinder or promote erosion. Episodes of soil erosion go back tens of thousands of years, whether induced by tectonic, climatic or human activity (as outlined in Chapter 3) and some soils have developed on the redeposited sediments of these erosion periods. In contrast, other soils have survived relatively intact from past periods and so may be regarded as relict soils or legacies of the past. Elsewhere soils are relatively young. All soils are, of course, being actively modified by present-day processes and in places this means being degraded. Soil degradation has even been recognised as desertification, a term more commonly thought of as applying to parts of Africa, south of the Sahara. To this end, research projects such as MEDALUS (Mediterranean Desertification and Land Use) have sought to establish the extent of the problem. A review of some of the MEDALUS findings forms the last part of the chapter.

4.1 Soil types

4.1.1 Soil classification

Soil scientists classify soils into categories generally according to soil properties, which are heavily dependent on the laboratory analysis of soil chemistry. Classifications such as those used by the United Nations Food and Agriculture Organisation introduce complex terminology and differ from one country to another. Not only is this confusing to non-specialists but it has resulted in the abandonment of common names such as red or brown Mediterranean soils. Objections to common (and more readily understood) names are, in part, due to the inappropriateness of applying a geographic term to soils which might be found beyond the boundaries of the geographical region. However, common names may be more useful in the Mediterranean region than those of formal soil classifications, especially as red soils and sediments are common in some parts of the Mediterranean and have an association with archaeological sites, as in Greece (see van Andel, 1998). Common names are, however, still in frequent use, not least in this book. The decision to use them here is based on a desire not to alienate those readers who are not soil scientists and because some common terms, such as terra rossa, are a part of the national soil classifications of countries such as Israel and Italy (Yaalon, 1997). Red soils have also been the subject of considerable research interest in regions with mediterranean-type climates (see, for example, a special issue of the journal *Catena*, volume 28, 1997) and this has promoted the use of the name, and by extension other common names.

In the Mediterranean and other areas with a mediterranean-type climate, the main soil types are brown Mediterranean soils, red Mediterranean soils, red brown earths, non-calcic soils and cinnamon soils (Bridges, 1978). Their

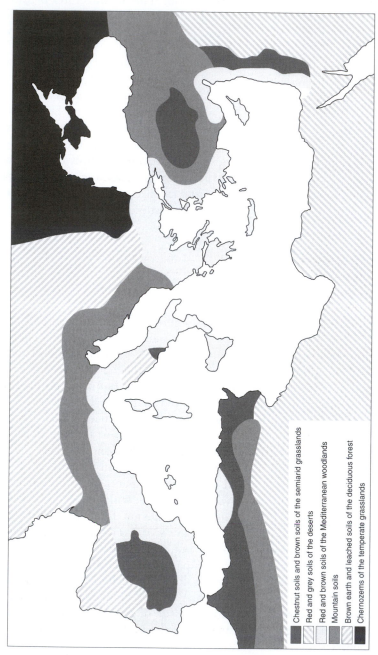

Figure 4.1 Mediterranean soil-type distribution map. After Bridges (1978).

Chestnut soils and brown soils of the semiarid grasslands

Red and grey soils of the deserts

Red and brown soils of the Mediterranean woodlands

Mountain soils

Brown earth and leached soils of the deciduous forest

Chernozems of the temperate grasslands

Table 4.1   Synonyms for soils in regions with a mediterranean-type climate

Common name	UNESCO/FAO World map	US Soil Taxonomy
Brown Mediterranean soils	Orthic luvisols	Hapludalfs
Red Mediterranean soils	Chromic luvisols	Rhodustalfs
Red/brown earth soils		
Non-calcic brown soils	Orthic luvisols	Haploxeralfs
Cinnamon soils	Chromic luvisols	Ustochrepts

Source: Bridges (1978).

distribution is shown in Figure 4.1 and, for the purposes of comparison, Table 4.1 gives some equivalents between these common names and those of the US Soil Taxonomy and the UNESCO/FAO World Map of soils.

4.1.2 What makes red soils red?

Colour is a major distinguishing factor of soil type, in particular the degree of redness. The colour comes not just from the amount of dark organic matter incorporated into a soil but also from the presence of oxidised iron. The amount of this is generally related to climate, but may also be a function of time and the rate at which iron is released on weathering (Birkeland, 1984). In terms of climate, the seasonal pattern of soil moisture – desiccation in summer and wetting in winter – and temperature are important. Worldwide, red soils are not common in regions of low temperature and varying amounts of precipitation. Rather, they are found in hot humid and hot arid environments (Birkeland, 1984).

As iron-bearing rocks weather, iron is released. Where this comes into contact with oxygen, as in oxygenated waters or a well-aerated soil environment, the iron is oxidised to form an iron oxide (ferric iron, iron III) such as haematite. This gives the pigment responsible for the red colour of soils which develops in conditions of relatively low water activity, high temperature, good aeration and/or high turnover rates of organic matter (Boero and Schwertmann, 1989; Birkeland, 1984). The pigment dominates soil colours even when the iron oxide content is less than 3%. Colour is generally irreversible, and indeed the degree of redness increases with the degree of weathering and hence age of the soil (Birkeland 1984; Bech *et al.*, 1997). Haematite is even regarded as an indicator of the special soil environment in which terra rossa soils form (Boero and Schwertmann, 1989). In regions with mediterranean-type climates, wet winters and summer drought appear to be the important factors in determining colour. The chemical reactions of the iron-bearing minerals are complex but it is suggested that during the wet period relatively high amounts of iron are released from the limestone parent material through weathering, resulting in the formation of ferrihydrite. In the oxidising conditions of the summer soil environment this forms haematite (Boero and Schwertmann, 1989).

4.1.3 Terra rossa soils

Terra rossa soils form particularly on hard limestones with high permeability and a soil cover of low water storage compared with softer limestones or marls, which are less permeable and therefore do not provide the same oxidising environment (Boero and Schwertmann, 1989). In addition such environments have a neutral pH which promotes oxidation. In acidic and anoxic, or anaerobic, conditions iron is reduced (ferrous iron or iron II – this has a blueish-grey colour). The combination of hard limestone and dry soils was shown to be important in the formation of terra rossa soils by Boero and Schwertmann (1989) in a wide survey of such soils from Australia, France, Germany, Greece, Israel, Italy, Lebanon, Mexico, Spain and the USA. This showed that terra rossa soils were not formed, even on hard limestone, when soil water activity was high, nor were they formed on softer limestones.

In the Mediterranean region, terra rossa soils are found in Spain, southern France, Italy, the area of the former Yugoslavia, Bulgaria, Greece and Turkey. In general these soils are less than 100 cm deep, with a dark red, clay-enriched B horizon (Figure 4.2). They are rarely continuous over large areas, instead forming in depressions surrounded by limestone outcrops. Percolation of calcium-rich solutions from the surrounding limestone means that they may be more calcareous than expected, given their derivation through solution weathering of the substrate.

In general it is difficult to find undisturbed terra rossa profiles because of erosion and redeposition of limestone soils which have prevailed for thousands of years, including periods of climatic change. This has complicated the debate about their origins. An objection to limestone as the immediate source of terra rossa material stems from the fact that its non-calcareous residue is often too low to explain the thickness of soil that has accumulated. This argument has a long history – suggestions have been made that soils found on the limestone plateaus of Derbyshire, England, are composed largely of loess mixed with the insoluble limestone residue (Piggot, 1962). Similarly, there are suggestions of an additional input of iron-rich, aeolian material to the insoluble residue of Greek limestones. Many Mediterranean limestones are nearly pure carbonate with a very low content of silicate minerals, the insoluble residue (MacLeod, 1980). The mean insoluble residue for the Pantokraton limestones of Epirus in Greece is only 1.5%. From this, calculations show that about 130 m of limestone need to be weathered to produce the average soil depth of 40 cm. This is in line with Yaalon (1997), who cites an expected accumulation of only 1–8 mm of terra rossa soil material during the 10 000 years of the Holocene, given annual dissolution of only 10–40 μm, or 10–40 cm of limestone over that period. Further calculations suggest that for many Mediterranean limestones, dissolution of 50 m is needed to accumulate 50 cm of soil, and that this would require some 2 000 000 years ($\pm$ 50%) – too long for many existing soil depths (Yaalon, 1997). MacLeod also found that there was no correspondence between the particle size distribution of the Epirus terra rossa soils overlying the limestone and the insoluble residue from it, nor

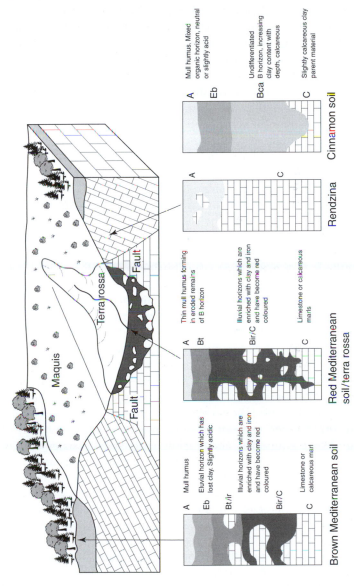

Figure 4.2 The relationship between soil types and topography. After Bridges (1978) and Faulkner and Hill (1997).

The following labels appear within the figure:

Maquis

Terra rossa

Fault

Fault

Brown Mediterranean soil

A — Mull humus

Eb — Eluvial horizon which has lost clay. Slightly acidic

Bt/ir — Illuvial horizons which are enriched with clay and iron and have become red coloured

Bir/C

C — Limestone or calcareous marl

Red Mediterranean soil/terra rossa

A — Thin mull humus forming in eroded remains of B horizon

Bt

Bir/C — Illuvial horizons which are enriched with clay and iron and have become red coloured

C — Limestone or calcareous marls

Rendzina

A

C

Cinnamon soil

A — Mull humus. Mixed organic horizon, neutral or slightly acid

Eb

Bca — Undifferentiated B horizon, increasing clay content with depth, calcareous

C — Slightly calcareous clay parent material

between the mean iron content of the soil and that of the residue. He noted a high proportion of soil in the 2–60 µm (silt) size fraction. This particle size is typical of aeolian material. One potential source, loess from northern Europe, was discounted given its predominant particle size of 20–50 µm. In contrast, the median particle size of the soils was within the size range of Saharan dusts.

Various studies have demonstrated that dust-laden winds from North Africa blow across the eastern Mediterranean. Chester *et al.* (1977) identified dust in traps south and west of Cyprus which originated in the deserts of North Africa and the Middle East. The dominant clay minerals were kaolinite and illite, with chlorite and smectite present. MacLeod (1980) compared this composition with the Epirus soils, and found that the main difference was that the soils were richer than the dust in kaolinite, had less illite and contained significant amounts of vermiculite. He accounted for this by suggesting differences in the weathering of the limestone and development of the soils. Through leaching, illite, chlorite and montmorillonite break down to form vermiculite, and kaolinite forms with desilification, at the expense of other clay minerals. In a later study, Nihlén and Mattson (1989) analysed meteorological data from a number of Greek weather stations and identified a significant dust component, originating from the Sahara, with maximum inflow in spring. They calculated that, with a 'guestimate' of an annual 11.4 g/m^2 extrapolated over the Holocene (10 000 years), 8.8 cm of dust would accumulate.

The hypothesis of a major dust contribution to soil formation was tested further by Pye (1992). Through laboratory analysis he compared red soils, surface sediments, bedrock samples and aeolian dust, and concluded on the balance of available evidence that it is unlikely that more than 50%, and perhaps less than 20%, of the present soil material is of aeolian dust origin. A more important source of parent material is provided by the weathering of local rocks. A much older contribution of dust has also been suggested (King *et al.*, 1997). Wind-blown sediments would have accumulated in marine deposits contemporaneously with the formation of the limestones in the Tethys Basin (see Chapter 3). These would then have been released as the limestone was uplifted and became exposed to subaerial weathering. Despite all the research, the debate continues, and Yaalon emphatically states that we can 'conclude that practically all terrestrial soils in the Mediterranean region were affected by the addition of eolian (*sic*) dust, adding significant amounts of fine soil material. This activity started some 5 000 000 years ago when the Sahara became a desert' (Yaalon, 1997, p. 163). Such a contribution would not be confined solely to terra rossa soils but would be found in all soils developing in the region. There would also be uneven spatial distribution of dust according to distance from source area, prevailing winds, slope, aspect and vegetation cover.

4.1.4 Brown Mediterranean soils

Where the period of summer drought is short and winter rains are more extensive, soils tend to be brown because leaching predominates; hence the traditional notion of brown Mediterranean soils (Bridges, 1978). Such soils develop on both non-calcareous and decalcified, originally calcareous, rock types. Leaching leads to an enrichment of clays in the B horizon (see Figure 4.2). In general, the A horizon is friable and rich in humus, while the B horizon is denser and less friable. Clays are also deposited on the linings of cracks in the C horizon as the soil dries in the hot summers. Desiccation also leads to the irreversible transformation of ferrous iron (iron II) to ferric iron (iron III) – see Section 4.1.2 above. This imparts a reddish colour to the lower horizons of these soils (Bridges, 1978). Brown Mediterranean soils are more extensive in the northern Mediterranean where they are a continuation of the brown soils formed in cooler, wetter climates.

4.1.5 Other soil types

In regions where drought lasts longer (five to six months) the colour of soils changes, giving cinnamon (yellowish-brown) soils characteristic of the drier Mediterranean climates (Bridges, 1978). This is the result of calcification of the soil – a characteristic of low-rainfall regions. Leaching is only slight as wetting of the soil by rain is confined to the upper layer (1–1.5 m) and soil moisture evaporates quickly. Consequently clay content increases with depth and calcium carbonate accumulates in the B horizon or the upper C horizon (see Figure 4.2). The A horizon is moderately rich in organic matter, 4–7% (Bridges, 1978). In the Mediterranean, such soils are found in eastern Spain, the Balkans, Turkey and North Africa. These are the drier regions of the basin, transitional to the even drier and hotter climates of the continental interiors.

Brown, red and cinnamon Mediterranean soils are described as zonal soils – those whose distribution reflects climatic rather than topographical or geological factors. In contrast, intrazonal soils have characteristics that are attributed more to local factors such as relief, drainage and parent material (Ellis and Mellor, 1995). Rendzina soils form mainly on limestone outcrops with dark, organic-rich, fine-textured, crumb-structured and shallow profiles (Bridges, 1978). They are poorly developed, with an A/C profile (Figure 4.2). The B horizon does not develop because weathering products are removed from the profile in solution. Similar shallow and poorly developed soils with no B horizon include rankers. The differentiation of their horizons is poor because of the limited time in which they have had to develop. They are associated with higher, steeper mountainous areas, where erosion also inhibits the accumulation of soil-forming material. Lithosols develop on weakly weathered bedrock, often on steep slopes, and are subjected to high water stress and are prone to erosion. As their names suggest, calcaric lithosols develop on highly permeable limestone, while eutric lithosols develop on non-calcareous parent material (Yassoglou, 1998). Regosols develop on a weathered mantle of rock debris. In some areas hard layers of calcium carbonate occur. These are calcrete soils, known as *croûte calcaire* in Algeria or *kafkalla* in Cyprus. Calcium carbonate accumulates within the soil profile but, through erosion of overlying material, becomes exposed at the surface, and then hardens to form impermeable layers.

Alluvial soils form in areas which are poorly drained. Some of these are agriculturally important soils reclaimed through wetland drainage. For example, the 350 km² Kopais basin in central Greece was finally drained, after attempts dating back to the Mycenaean period some 3500 years ago, in the late nineteenth and early twentieth centuries (Allen, 1990, 1997). Today it is a fertile, important agricultural region.

4.1.6 Are soils typical of the Mediterranean unique to the region or widely found elsewhere?

Soils typical of the Mediterranean Basin are not confined solely to that region and are found in other regions with a mediterranean-type climate or similar lithological characteristics (Bridges, 1978). This should not be surprising given that soil properties are a function of time, parent material and climate. These are not uniform throughout the Mediterranean Basin and nor are they restricted to it. Red soils, for example, have been identified well away from areas with mediterranean climates (Birkeland, 1984). However, sometimes these are palaeosols which formed in periods of more arid and warmer climates and which have survived to the present. Local occurrences of relic terra rossa type soils are known from Wexford in Ireland (Gardiner and Ryan, 1962). These are found buried beneath glacial drift and it is suggested that they developed on the underlying limestone conglomerate in a warm climate. With the subsequent advance of ice, much of the terra rossa soil would have been eroded, except from small protected depressions where it now remains as a truncated buried soil.

4.2 The role of time in Mediterranean soil development – mineral matter and organic matter

Unlike soils of northern Europe which were actively modified during the cold stages of the Quaternary, those of the Mediterranean were less affected. They should therefore be better developed, with greater differentiation of soil profiles and a mineralogy reflecting a long period of subtropical climate. The result is that the mineral part of many soils and the organisation of their horizons should have the characteristics of palaeosols (soils formed in an environment of the past). For example, at least part of the redness of soils is a function of time (Birkeland, 1984; Bech et al., 1997) and of the relatively continuous period of soil oxidising conditions, first in the arid climates of the glacial stages and then in the interglacial periods, including the Holocene. However, the organic component of Mediterranean soils is more modern, influenced by recent and present climatic conditions (Bottner et al., 1995). This distinction becomes important because soil erosion related to human activity affects essentially the upper horizons where the organic matter accumulates. Rates of formation of organic matter are then important in any questions relating to the renewability or sustainability of soils (see Section 4.3).

The amount of organic matter in Mediterranean soils is variable (Bottner *et al.*, 1995). It is closely controlled by the soil mineral environment and therefore parent material. Soils which develop on calcareous parent material tend to have high amounts of organic matter which is deeply distributed through the soil. This occurs because the clay minerals remain in the topsoil and high concentrations of Ca^{2+} ions protect the organic matter against microbial decomposition. In some such soils organic matter may remain in the soil for up to 3000 years. In contrast, soils which are leached of clays and carbonates have decreasing organic carbon contents and shallower organic A horizons. The turnover time and mean age of the organic matter in these soils is commensurately reduced. As soils become more leached and acidic, so most of the fine mineral fractions are removed from the upper horizons; only sandy material remains in the topsoil. This lowers the water retaining capacity of the soil and also reduces the activity of earthworms. They are essential for the incorporation of organic matter into the mineral matter in the topsoil. Reduced earthworm and soil faunal activity means reduced decomposition of litter and the formation of a shallow, raw humus layer. Comparisons of organic matter content in soils formed on different substrates, but all with a forest cover, show highest content on limestone (mean values 115.7±26.53 mg per hectare) and lowest on calcareous sandstone (71.3±26.06 mg per hectare).

As well as parent material, the amount of organic matter in soils is a function of climate. Summer drought promotes deep root systems for many perennials seeking soil water (see Section 5.4) and high summer temperatures favour the deep below-ground activity of soil fauna. Both of these factors contribute appreciable amounts of soil organic matter (Bottner *et al.*, 1995). The Mediterranean climate also influences some soil chemical processes such as moderately acidic to basic soil pH, which further promotes biological activity. A consequence of this climatic effect is that global warming would be expected to affect soil organic matter, both directly through modifying decomposition rates because of higher soil temperatures, and indirectly through modifications to the quality of soil litter in response to increased levels of carbon dioxide. As soil organic matter is more readily eroded than mineral matter any change in its content has significance for future soil losses.

4.3 Soil erosion

Soil erosion in the Mediterranean region has a long history. Soils have been identified that are associated with Late Pleistocene erosion and deposition of the 'older fill' (see Section 3.4). It is generally agreed that climatic change was the principal mechanism; a cooler and drier climate, with sparse vegetation cover, caused increased sediment yield from catchments. The debate concerning erosion history during the Holocene is more complicated, with respect to both timing and causal factors. At times, the argument has been polarised between those favouring a climatic cause (for example, Vita-Finzi, 1969, 1976; Bintliff, 1977) and those favouring an anthropogenic cause (for example, Wagstaff, 1981; Chester and James, 1991). Even when land use is known, there is no consensus as to the effects of change in land use; there is considerable discussion about the role of terraces in erosion control, and whether abandonment of terraced cultivation results in increased erosion (McNeill, 1992) or whether terraces continue to protect slopes once abandoned. As terraced cultivation ceases, vegetation cover changes. Rackham and Moody (1996) note the prevalence of pine woods on derelict Cretan terraces, and the subsequent risk of erosion following wildfires in pine forests. Thus the 'story' of erosion is by no means simple: 'The variety of erosion in the Mediterranean defies generalisation. Erosion in Crete is not the same as in Epirus' (Rackham and Moody, 1996, p. 22). Arguments about whether Mediterranean landscapes and soils are resilient or fragile are therefore superficial. The terms resilient and fragile, in themselves, are also somewhat obtuse; they need to be defined with respect to particular land uses and time scales.

Questions of resilience and fragility often involve the notion of renewability or sustainability and so need to be based firmly in a temporal framework. If soils are regarded as renewable then do we need to be worried about soil erosion or depletion? If soils are regarded as non-renewable, then they need to be cherished (Trudgill, 2000). As most soils are a legacy of past processes but are being actively modified by human activity, Trudgill argues that it is foolish to regard them as renewable. Even if weathering keeps pace with erosion, this only applies to the formation of mineral matter, whereas erosion is generally that of organic matter. As external inputs of organic matter would be needed to supplement losses, many soil uses are best regarded as unsustainable.

4.3.1 Long-term perspectives

The landscape history of Mediterranean Europe has traditionally been viewed in terms of degradation and desertification from a golden age in the distant past (Rackham and Moody, 1996) – the 'ruined landscape theory' or the idea that in classical times, and indeed into the historic past, Mediterranean people lived in a lush, wooded landscape, not unlike that of northern Europe. The actions, especially through war and misgovernment, of the Romans, Arabs, Venetians, Turks and Ottomans were regarded as culminating in the degraded landscape described by travellers of the eighteenth century and onwards. Descriptions were placed firmly within the context of an appreciation of the landscapes of cooler regions, not those of a semiarid area. Such views might have been reinforced by the reading of classical authors such as Plato (fourth century BC): 'Consequently, since many great convulsions took place during the 9000 years . . . the soil which has kept breaking away from the high lands during these ages and these disasters, forms no pile of sediment worth mentioning, as in other regions, but keeps sliding away ceaselessly and disappearing in the deep. And just as happen in small islands, what now remains compared with what then existed is like the skeleton of a sick man, all the fat and soft earth having wasted away, and only the bare framework of the land being left' (Plato, 1929, p. 111). Yet, Rackham and Moody (1996) argue that this passage is a work of fiction and should not be taken literally. However, the ruined landscape theory still exerts a pervasive, if unacknowledged, influence on writers about the Mediterranean, but of course in reality the theory is too simplistic and the relationships between the contemporary landscape and the past are far more complex; so too those of landscape and climate, both past and present.

4.3.1a Soils in antiquity

What can we learn about the effect that people in antiquity had on their soils for comparison with the present day? A partial answer is provided by Davidson (1980), who examined soils on Thera, in the Aegean, buried beneath pumice and tephra from the explosion of Santorini. The dating for this is problematic, but is likely to have been c. 1630 BC (Hardy and Renfrew, 1990). Davidson measured particle size distribution, organic content and cation exchange capacity for six Minoan (pre-eruption) soils and compared these with two modern soils. Results showed similarities between the two samples, and the soil characteristics of both strongly suggested that they were the products of accelerated erosion. None of the exposed Minoan soils showed any development of soil horizons. The soils were coarse – predominantly sand with only small proportions of silt, and almost no clay. Organic carbon content and cation exchange capacity were low. Such characteristics are consistent with erosion and deterioration caused by the Minoans rather than those which would have been expected in the absence of human and volcanic activity during the period of about 15 000 years which preceded the Minoan eruption of Santorini. Present soils on the island are being actively eroded and are poorly developed, despite 3500 years of pedogenesis since the eruption. The value of this study lies in its incontrovertible dating – anthropogenically induced, accelerated soil erosion was taking place at least 3000 years ago, in parts of the Mediterranean Basin.

4.3.1b Terraced agriculture and soil erosion

In many countries of the Mediterranean, terraced agriculture has been a typical element of the landscape (Plate 4.1), and features in any speculation about land use and soil erosion. A good example comes from the historical phases of

Plate 4.1   Terraced agriculture in the serra (uplands) of Monchique, Portugal.

soil erosion associated with the role of terraced agriculture in Alpujarra, Spain. McNeill (1992) identified four periods when neglect of terraces resulted in erosion. These periods reflect the political history of the Iberian Peninsula, and are all within the last 500 years. The first came after the *Reconquista* (reconquest) and expulsion of the Moors in the fifteenth century AD. The Moorish settlers had established a system of well-tended, terraced and irrigated agriculture. With their expulsion, new settlers, unfamiliar with the local environment, introduced new farming methods – unterraced and unirrigated – which, it is argued, led to soil erosion (Douglas *et al.*, 1994). Expulsion of the Moors led to such transformations of the countryside that whole settlements were washed away, with the greater part of the region's topsoil washed down to the deltas of the Mobril and Adra rivers (McNeill, 1992). The second phase of erosion in the late eighteenth and early nineteenth centuries occurred with the extension of mining and market agriculture into the region. This overlapped with the third phase in the mid-nineteenth century as a result of deforestation and further extension of agriculture. Finally, erosion rates are believed by McNeill to have increased since the Second World War with population decline, agricultural deintensification and abandonment of terraces (but see below for an alternative argument). There is, however, uncertainty about present erosion rates in Alpujarra. Douglas *et al.* (1994) cite a variety of sources for erosion rates, which give figures of between 370–575 $t/km^2/yr$, up to 2000 $t/km^2/yr$ in the high Alpujarra, and more than 20 000 $t/km^2/yr$ in some highly localised, gullied areas. Obviously, there is a high degree of spatial variation in sediment yield. It is therefore difficult to extrapolate or generalise about erosion rates – either today or, even more so, in the past.

In broad contrast to McNeill's view that agricultural abandonment in the form of terrace neglect leads to episodes of erosion, Thornes (1998b) holds a largely positive view that abandonment of agricultural land and revegetation are reducing runoff and erosion and that under the present dry climate, soil erosion rates, at least at the field scale, are significantly lower than has commonly been believed. There is also evidence from pollen records that abandonment of agriculture results in significant revegetation with a stabilising effect (Barbero *et al.*, 1990; see also Section 8.8.1). It is possible that much of the damage done by erosion occurs as a result of high magnitude events with a frequency longer than that of the average research project, rather than as a result of ordinary rainfall. Such an idea is based on research during the last decade which has challenged the notion of the Mediterranean region as one of fragility, subject to and at risk from high rates of soil erosion. These themes are now examined further.

4.3.2 Contemporary issues in soil erosion

In 1984, a symposium of the Commission of the European Communities on Desertification in Europe, met at Mytilene in Greece (Fantechi and Margaris, 1986). At the time, desertification was not regarded as a European problem, and the idea that it should be was indeed questioned at the symposium.

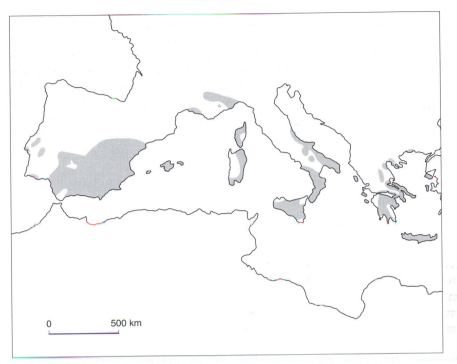

Figure 4.3 Map showing areas vulnerable to desertification in the northern Mediterranean. After Yassoglou (1998).

Today, European desertification is still not widely recognised or accepted (Middleton and Thomas, 1994) and a better description of the 'desertification' problems faced in Mediterranean countries is probably environmental degradation (Grove, 1986). This is 'the culmination of centuries of environmental change under difficult social, political, economic and physical conditions of Mediterranean Europe' (Mairota *et al.*, 1998, p. vii), though it would also apply to countries of the Middle East and North Africa. Areas of Europe vulnerable to desertification are shown in Figure 4.3. Following the symposium, the European Commission funded the MEDALUS (Mediterranean Desertification and Land Use) Project, with researchers from eleven different nations and many different disciplines examining the physical, biological and human aspects of environmental change, taking place currently and in the past, within the region.

Under the MEDALUS project, study sites were established in Portugal, Spain, Italy and Greece. Erosion plot measurements were made, and the data were used for calibrating models of slope response to storm runoff and water erosion. Three component models have been used: SHETRAN, a widely used model for routing water and sediment through catchments of 5–500 km², MEDRUSH, which simulates runoff and erosion in catchments of up to 2500 km² for periods of up to 100 years, and the MEDALUS model, which simulates changes down the length of a single hillside flow strip or catena

(Kirkby *et al.*, 1998). The understanding of catenary processes gained from the MEDALUS model is applied in the MEDRUSH model to representative slope catenas within each of several hundred subcatchments which comprise the study area. Various initial parameters, such as seasonal and longer term changes in vegetation cover and soil conditions, are altered in the model to provide dynamic predictions of runoff and erosion. In consequence, the MEDRUSH model simulates some of the factors and processes which may lead to irreversible desertification, or reversible soil degradation. It is intended that the results of these models, nested one within another, can be of use to planning authorities.

In order for the outputs of the models to have validity, direct knowledge of the interaction between all the component parts of the Mediterranean biosphere is needed. Thus, research within the framework of the MEDALUS project has included many themes – palynological studies to determine past vegetation patterns, examining timing, natural trends and human impacts; modelling of future climates; assessment of remote sensing for monitoring the extent and progression of desertification; remote sensing of land degradation; risk assessment and land-use planning; modelling of vegetation dynamics and degradation of ecosystems; hydrological responses to land-use changes and examination of the atmospheric boundary layer, surface energy budgets and soil–vegetation–atmosphere transfer. In consequence, the literature related to climate, anthropogenic history, land use, erosion and allied factors for the Mediterranean region was prolific during the 1990s. Some of the conclusions will be discussed in Chapters 8 and 9 on land use and environmental issues. There are, however, some results that throw light on the question of the fragility or resilience of Mediterranean soils and their erodibility. Two such issues are considered further: the effect of vegetation cover on runoff and soil erosion rates, and the role of rock fragments in protecting soils from erosion.

4.3.2a Vegetation cover, annual precipitation and sediment loss

It has long been recognised that vegetation cover affects runoff and soil erosion, and high erosion rates (as measured by sediment yield) are generally expected in regions with a mediterranean-type climate (see Section 3.4). With a long history of human settlement and a variety of substrates, the Mediterranean Basin provides a good laboratory for examining the relationship between vegetation (land use) and precipitation, and annual runoff and sediment loss (Kosmos *et al.*, 1997). As part of the MEDALUS project, eight experimental sites were located in Portugal, Spain, France, Italy and Greece representing a variety of landscapes and lithologies. Runoff was monitored under the following land uses: olive trees, shrubland, rain-fed cereals, eucalyptus and vines. Despite the small number of plots and large standard deviations for the measurements on some land-use types, the results showed an explicable variation in average sediment loss according to vegetation cover (Table 4.2).

Olive trees tend to be grown in semi-natural conditions, with annual vegetation growing between the trees, and little removal of plant residue from the soil surface during the year. This prevents surface sealing of the soils, and

Table 4.2 Average sediment loss (and standard deviations) for different land-use categories from sites in Portugal, Spain, France, Italy and Greece

Land use	Average sediment loss ($t\,km^{-2}y^{-1}$)	Standard deviation	No. of plots
Olives	0.8	3.3	3
Shrubland	6.7	5.6	95
Rain-fed cereals	17.6	26.1	65
Eucalyptus	23.8	34.2	12
Vines	142.8	157.5	9

Source: Kosmos *et al.* (1997).

reduces the velocity of surface runoff. The result is reduction of soil loss to a minimum. Olives thus protect hilly areas from soil erosion, even where these grow on abandoned terraces (see Section 4.3.1b). However, the economics of olive growing are changing. European Community subsidies have led to increases in the area of olive cultivation. Rackham and Moody (1996) describe most Cretan valleys as being a 'sea of olives from cliff to cliff [where] the point has been reached that enough is enough' (p. 211). New terraces have been built to grow more olives, but terraces today are being built without stone retaining walls. It remains to be seen whether the increased olive cultivation will result in greater protective cover for soils or in further erosion.

The relationship between annual precipitation and sediment loss under shrubland shows a trend of increasing loss with decreasing precipitation, when precipitation is greater than 280–300 mm per year. Where precipitation is below this figure, sediment loss decreases with increasing aridity. Greatest sediment loss in the MEDALUS study occurred at Almería, southern Spain, under an annual precipitation of 282 mm per year. A comparable relationship is evident between annual precipitation and runoff. Peak runoff occurs under an annual rainfall of 280–300 mm. Below this, runoff decreases with aridity, and above this runoff decreases with increasing precipitation.

Rain-fed cereals, such as winter wheat, do not afford protection to soils from winter rainfall (early October to late February) as the soils are almost bare or the vegetation cover is not sufficient to protect the soil from the high-intensity, long-duration rainfall. Hence sediment loss is greater than under shrubland or olives. It also increases as annual precipitation increases, especially above 380 mm per annum. One mitigating factor in reducing sediment loss under winter wheat is the presence of rock fragments in many of these cultivated soils. As described below (Section 4.3.2b), rock fragments tend to protect soils from erosion; without them sediment losses were expected to have been greater than actually measured. Sediment losses increase further under stands of eucalyptus. Soils beneath the canopy are almost bare as eucalyptus competes for water with other vegetation, especially in stands of young trees, hence runoff and sediment loss increase. This is significant Basin-wide, as eucalyptus is a widely planted exotic tree (see Sections 4.5 and 8.9). The

4.6 Soils, ecosystems and soil conservation

The properties of soils are fundamental to the way in which ecosystem processes operate, helping to regulate the flows of energy and the biogeochemical cycles – the cycling of gases, non-gaseous compounds and water. Soils are also fundamental to plant growth and therefore the animals which inhabit the vegetation communities. For continued maintenance of these ecosystems some forms of soil conservation are needed. These may be terraces dating back thousands of years. Other conservation measures include discarding stones and rocks into gullies to prevent any further downcutting of the gullies, the construction of small check-dams to hold back soil and water, and planting of protective vegetation cover (Conacher and Sala, 1998). These issues of soils and ecosystems are examined further in Chapter 7.

4.7 Further reading

Birkeland, P.W. (1984) *Geomorphology and Soils*, Oxford University Press, Oxford.
Ellis, S. and Mellor, A. (1995) *Soils and Environment*, Routledge, London.
Trudgill, S.T. (2000) *Terrestrial Biosphere*, Pearson Education, Harlow.

Chapter 5

Plants and animals

'Having travelled about two hours into the Valley, we entered into a Woody Mountainous Country, which ends the Bashalick of Aleppo, and begins that of Tripoli. Our Road here was very Rocky, and uneven; but yet the variety, which it afforded, made some amends for that inconvenience. Sometimes it lead us under the cool shade of thick trees: sometimes thro' narrow Valleys, water'd with fresh murmmuring [*sic*] Torrents: and then for a good while together upon the brink of a Precipice. And in all places it treated us with the prospect of Plants, and Flowers of diverse kinds: as Myrtles, Oleanders, Cyclamens, Anemonies, Tulips, Marygolds, and several other sorts of Aromatic Herbs.'

Henry Maundrell (1749, diary entry for Monday, 1 March 1697)

Henry Maundrell's diary provides a sense of place for the eastern Mediterranean with many references to its landscape and vegetation communities. One objective of ecogeography is to explain these communities. Plant species richness in the Mediterranean region is high, with an estimated 25 000 flowering plants and ferns in an area of about 2.3 million km², compared with just 6000 species in non-Mediterranean Europe, an area of 9 million km² (Quézel, 1985). However, compared with other areas of the world which have a mediterranean-type climate species richness is not as high, though two particular centres of species richness, including high rates of endemism, occur – in the region of Morocco and the Iberian Peninsula combined, and in Turkey and Greece. Distributions such as these reflect ecological processes operating at varying temporal and spatial scales, from the environmental factors operating over evolutionary time to small-scale interactions between organisms and their physical and biotic environments, such as the adaptations of plants and animals to climate and fire.

5.1 Explanations for distribution patterns

5.1.1 Long-term perspectives

Maps showing distributions of plants and animals across the world reveal that some genera and species are found in several separated localities. These are known as disjunct distributions; an example is the generic affinities between

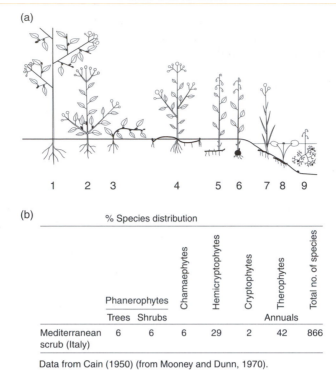

Data from Cain (1950) (from Mooney and Dunn, 1970).

Figure 5.1 Raunkiaer's life-forms and frequencies for the Mediterranean region.
Source: Goudie *et al.* (1994). (a) Raunkiaer's life-forms: (1) Phanerophytes, (2,3) Chamaephytes,
(4) Hemicryptophytes, (5–9) Cryptophytes. The parts of the plants that die during the
unfavourable season are shown unshaded, the persistent portions and perannating buds are
black. (b) Proportions of Raunkiaer's life-forms in scrub plants of Mediterranean Italy.

the plants in the Mediterranean region itself and in mediterranean California
(see Section 5.2.2). Early historical biogeography explanations for such distributions were based on dispersalist hypotheses, which argued that taxa could
migrate across barriers from one location to others. In other words, the ability
of a taxon to spread and its method of dispersal were the important factors in
determining its distribution. Long-distance dispersal was believed to be possible
across barriers such as deserts, mountains and seas. More recent explanations
have centred on vicariance biogeography which emphasises the erection of
physical barriers separating populations already in existence into two or more
isolated populations. Mechanisms of isolation include both continental drift
and long-term climate change. As continents separated and drifted apart, or
collided or were subducted to create mountains, so oceanic and mountain
barriers formed. Long-term climatic changes might have resulted in the development of isolating barriers such as deserts. Once disjunct populations have
become established, speciation within these divided populations may lead to
the development of distinct, new populations, or species, for example the different species within the shared genera of California and the Mediterranean.

Although the dispersalist and vicariant approaches have been viewed as fundamentally opposing, in reality both types of processes operate and contribute to the complexities of animal and plant patterns (Stace, 1989).

5.1.2 Ecological biogeography

As well as adopting a historical approach, interactions between organisms and their physical and biotic environments can further explain species distributions, focusing on mechanisms by which organisms are restricted in their range and those that maintain species diversity. Plants and animals exhibit structural and physiological adaptations to a range of environmental factors. These help to determine life-form – shape, structure, habit and life history. Life-form can be studied at the level of individual features such as the number of stomata in a leaf, or at the level of the whole organism, for example whether a plant is a tree, shrub or herb. A common, albeit dated, classification of plants in terms of life-form is that of Raunkiaer, based on the positioning of over-wintering buds, for example whether above or below ground. This is shown in Figure 5.1. Such a classification allows comparisons between life-forms in differing climatic regions and environmental habitats. Ecological responses of plants and animals to environmental stimuli can also aid our understanding of parallels in life-forms and also between community structures and ecosystems (see Chapters 6 and 7) in different regions of the world.

5.2 Historical development of the present Mediterranean flora

5.2.1 Explanations for Mediterranean plant distribution based on source area

The origins of a region's flora can be understood partly in terms of source area; genera derive from different biogeographical regions or floristic realms. Floristic realms are characterised by plant families, and substantial proportions of these families, that are endemic within that region. Several different maps have been produced to show these realms and one such is shown in Figure 5.2a (while Figure 5.2b shows the world's faunal realms – see Section 5.3). The main floral realms are the Holarctic, Neotropical, Palaeotropical, Antarctic, Australian and Cape Kingdoms. Their distribution reflects the earth's geologic history – the influence of continental drift set against the evolution of flowering plants. For the Mediterranean, identified as a separate floristic region within the Holarctic kingdom, a major element of the flora is tropical. This includes taxa which are shared with Africa and Asia (the Palaeotropics), the Cape Kingdom, South America (the Neotropics) and Australia. Examples in this group include the fig (*Ficus*) and the vine (*Vitis*). The wild progenitor of the cultivated fig (*Ficus carica*) is closely related to types of *Ficus* found to the south and east of the Mediterranean region and the genus itself comprises more than 2000 species mostly found in the tropics (Zohary and Hopf, 1993). The cultivated grape vine (*Vitis vinifera*), once thought to be an independent

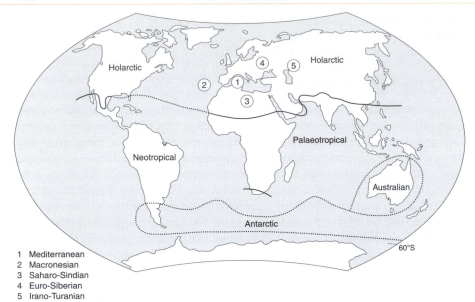

1 Mediterranean
2 Macronesian
3 Saharo-Sindian
4 Euro-Siberian
5 Irano-Turanian

Figure 5.2a Floral realms (or kingdoms) of the world showing selected floral regions. Adapted from Cox and Moore (1993) and Goudie *et al.* (1994).

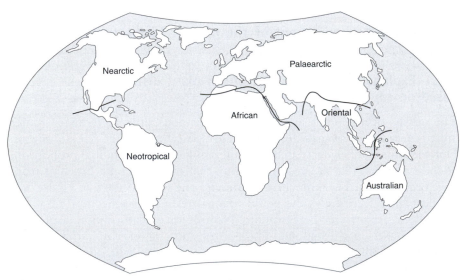

Figure 5.2b Faunal realms of the world. Adapted from Cox and Moore (1993).

species, is now believed to be derived from the wild grape (*V. sylvestris*). Confusingly, the latter is now classed as a subspecies of the cultivated grape – *V. vinifera* var. *sylvestris*. The range of this wild vine extends from the western Himalayas to the Atlantic. Other species of *Vitis* are found in central and eastern Asia and in North America (Zohary and Hopf, 1993). These examples of tropical taxa may be described as having a pantropical origin; that is, not confined to one tropical region. Certain taxa, however, are more restricted in their origin, such as those which are African (or Palaeotropical) in origin but may have been present in the circum-Mediterranean region since at least the Oligocene and Miocene. These include genera such as asparagus (*Asparagus*), carob (*Ceratonia*), palm (*Chamaerops*), jasmine (*Jasminum*), olive (*Olea*), oleander (*Nerium*) and *Phillyrea*. They are not found in the semiarid floras of the Americas, and Raven (1973) suggests that this is because they remained on the fringes of the semiarid regions of tropics of the Old World for a long time before moving towards the Mediterranean region sometime after the splitting of the North Atlantic.

5.2.2 Explanations for Mediterranean plant distribution based on climate

Many elements of the Mediterranean flora can be described as climatically 'mediterranean' in origin. Their development is associated with the beginnings of a semiarid, mediterranean-style climate, the earliest form of which is recognised as having occurred as early as the Cretaceous (Raven, 1973). These taxa evolved in areas bordering drier regions dominated by more sclerophyllous taxa. Such differentiation led to the appearance of mediterranean floras, adapted to semiarid environments. Included here is a common element in several genera of woody plants in the Mediterranean Basin and in California. For example, several representatives of the following genera are in this category: *Aesculus* (chestnut), *Arbutus* (strawberry tree), *Fraxinus* (ash), *Juniperus* (juniper), *Prunus* (cherries, peaches and plums), *Rosa* (rose), *Pinus* (pine) and *Quercus* (oak) (Raven, 1973). However, the species of these genera found in California and the Mediterranean are not the same. This implies a common origin at the generic level prior to the widening of the Atlantic Ocean. Speciation occurred once the Eurasian and North African plates separated from North America some 50 million years ago.

5.2.3 Explanations for Mediterranean plant distribution based on timing of origins

The origins of Mediterranean flora can be described temporally as well as spatially, with distinctions made between an older (palaeo-Mediterranean) flora and a more recent (neo-Mediterranean) flora. Palaeo-Mediterranean flora include those taxa for which there is fossil evidence before the Plio-Pleistocene. In other words, they evolved before the development of a 'mediterranean' climate in the mid- to late Pliocene, some 3.2 million years ago. Also included are those elements described above which evolved outside but migrated to the Mediterranean region prior to the break-up of the Eurasian, North African and North American plates, some 50 million years ago.

The neo-Mediterranean flora contain taxa absent from the pre-Pliocene record – they developed in a mediterranean-type climate. A recognisable 'modern' mediterranean-type climate in the Mediterranean region, with a rhythm of summer drought and winter rainfall, first appeared approximately 3.2 million years ago during the mid- to late Pliocene. This was the result of a global cooling, the establishment of permanent ice in the Arctic and the intensification of the North Atlantic Drift (Suc, 1984). Pollen evidence from offshore deposits in the Gulf of Lion and the Adriatic Sea, and from terrestrial sequences from Languedoc and Roussillon in France, and Catalonia in Spain consistently show a replacement of dense forest dominated by one of the conifer families, Taxodiaceae, with sclerophyllous forests and higher frequencies of xerophytic taxa such as olive (*Olea*), evergreen oak (*Quercus ilex*-type), *Phillyrea, Cistus* and *Pistacia*. Today these are all typical Mediterranean flora. It would, however, be wrong to assume stable climatic and vegetation patterns in the Mediterranean region since then.

5.2.4 Climatic fluctuations and the development of Mediterranean flora

The Pleistocene glacial/interglacial stages of higher latitudes are associated with marked climatic fluctuations in the Mediterranean region (see Section 2.7). Pollen evidence shows clear fluctuations between open vegetation, steppe communities during cooler periods and deciduous forest during warmer periods. The steppe communities were dominated by *Artemisia* (wormwood), Amaranthaceae and Chenopodiaceae (the goosefoot family). An example from around Lake Ioannina in Greece shows the deciduous forest communities to have been dominated by taxa such as oak (*Quercus*), elm (*Ulmus*), hornbeam (*Carpinus betulus*), *Ostrya carpinifolia*, oriental hornbeam (*Carpinus orientalis*) and fir (*Abies*) (Tzedakis, 1993). The longest pollen records for southern Europe come from the Massif Central in France (430 000 years), the Vale di Castiglione in Italy (250 000 years), and Ioannina (430 000 years) and Tenaghi Philippon (700 000 years) in Greece (Tzedakis *et al.*, 1997). These show that similar climatic fluctuations occurred across the northern shores of the Mediterranean Basin. Recent research also reveals that high-resolution climatic changes, as recorded in Greenland ice cores, are apparent in the vegetation history of the western Mediterranean region (Adams *et al.*, 1999) and the Adriatic (Asioli *et al.*, 1999). For the eastern Mediterranean, the existence of a vegetation response to cooling in the Younger Dryas (11 000 to 10 000 years BP) is equivocal (Bottema, 1995).

Although it seems that cold periods were dominated by steppe vegetation, pollen sequences also record the continuous presence of tree pollen, such as oak (*Quercus*) and pine (*Pinus*) at some locations such as Ioannina in Greece (Tzedakis, 1993). There may be some over-representation of these genera given the long-distance dispersal abilities of their pollen, but their presence suggests a continued existence, even if in small communities. This demonstrates that during cold periods, some parts of southern Europe acted as refugia for tree species from northern Europe (Bennett *et al.*, 1991). As well as pollen evidence for refugia, genetic markers demonstrate their existence (Thompson,

1999). Using maternally inherited chloroplast DNA, patterns of genetic differentiation among the refuge areas and the spatial population structure of different species across Europe allow recolonisation routes to be reconstructed. Many tree species show higher levels of polymorphism (the existence of two or more genetically distinct forms of the same species) in the Mediterranean region than in the recolonised areas. Only a subset of the genetic variation present in refugia is present at higher latitudes.

The mountains of the Balkans are believed to have been a major refuge area. This region is not thought to have been widely glaciated during the last glacial stage (at its maximum 18 000 years ago), and July temperatures are estimated to have been only 5 °C cooler than at present. Combined with the high altitudinal range, these factors helped to provide the variety of environments necessary for the survival of temperate taxa (Willis, 1994). Other likely refugia were in the Alps and Italian mountains, as well as along large river valleys such as the Rhône (Blondel and Aronson, 1995). With sea levels at least 100 m lower during the last glacial stage (see Section 3.7), a greater expanse of alluvial deposits would have been exposed, providing potential habitat for more temperate tree species.

The continued existence of Mediterranean forests with these temperate taxa is important, for it seems likely that during the interglacial to glacial transitions in northern Europe, the temperate trees did not retreat southwards but degraded *in situ*. Their reappearance after glacial stages was only possible because of their continued existence in southern Europe through the cold stages. The southern European sites are therefore long-term refugia, and will need to remain so if cold climates re-assert themselves, or act as sites from which repopulation can take place if deforestation becomes a major problem.

5.2.5 Lusitanian flora

An interesting component of Mediterranean flora is the Lusitanian element, named after the Roman province of Lusitania in Portugal. A number of species have a distribution which goes beyond the Mediterranean Basin, north along the west coast of Europe to the west of Ireland. These include the strawberry tree (*Arbutus unedo*), St Dabeoc's heath (*Daboecia cantabrica*) and Dorset heath (*Erica ciliaris*). These taxa are believed to have spread north from the Mediterranean in early post-glacial times along the western seaboard of Europe, but were then separated into disjunct distributions by rising sea levels. *Arbutus unedo* grows today in the countries of the northern Mediterranean and in the Maghreb and Levant (Figure 5.3). It is absent from Atlantic France and does not recur until Ireland, where it is found in southern Kerry, County Cork and Sligo – a disjunct distribution even within Ireland (Sealy and Webb, 1950). It has sclerophyllous leaves (see below) and the tree flowers in autumn, producing fruit through the winter. These characteristics are typical of mediterranean plants, which respond to winter rains and summer drought. It does not tolerate frost; winter temperatures largely control its distribution to those areas where

Figure 5.3 The distribution of *Arbutus unedo* in western Europe and the Mediterranean region. Adapted from Sealy and Webb (1950).

mean January temperature is above 4.5 °C. Its continued survival in western Ireland is only possible because of the mild climate determined by the warm Atlantic Gulf Stream.

5.2.6 Introduced or exotic flora

Biogeographically, the Mediterranean is at a crossroads, with its native fauna and flora either originating *in situ* or arriving from northern Europe, western Asia and Africa. This makes it difficult to define a 'native' species (di Castri, 1991). Those species described above as palaeo- or neo-Mediterranean would be included because they evolved in the region either before or after a mediterranean-type climate had developed. Some species might also be regarded as native if they are naturalised invaders, in other words those which have come from other biogeographical or climatic regions but are able to reproduce successfully and spread of their own accord throughout the region. Some introduced species have become invasive and a nuisance.

Invasive species are found in all mediterranean-type ecosystems, sometimes causing severe ecological problems, as in the fynbos of the Cape Province, where invasive plants, many of which are from Australia, threaten the survival of native, endemic species (Cowling, 1992). Of course, not all non-native species are invasive and some are now regarded as typical of mediterranean-type flora. The prickly pear cactus (*Opuntia ficus-indica*) is sometimes taken to be one such species (di Castri, 1991). The history of these introduced species follows that of contacts between people as they migrate from one area to another and trade with far-distant places. Between the five regions with a mediterranean-type climate major introductions occurred with the European colonisation of Africa, Australia and the Americas. Before the opening of the Suez Canal in 1869 trade between Europe and Australia went via the Cape, offering many opportunities for deliberate and accidental introductions between the three continents (Fox, 1990). A similar story is true of trade from the Iberian Peninsula to Chile and California before the opening of the Panama Canal. As a result there are now many common species between the five regions, although the cultural differences between Anglo-Dutch colonisers and the Spanish have led to greater biotic differences between Chile–California and Australia–South Africa (di Castri, 1991).

What makes a plant a successful invader? Those which grow alongside traditionally managed crops tend to be the most successful – in other words, weeds (Le Floc'h, 1991). Two examples in the Mediterranean Basin are corncockle (*Agrostemma githago*) and knapweed (*Centaurea cyanus*). Weeds have a preference for disturbed ground, with high demands for nutrients and a dislike of competition. Not surprisingly, as agriculture spread from the Near East so too did plants, in particular those with a short growing season. The ability to set seed before being destroyed by ploughing or harvesting gives these plants a competitive advantage. Examples associated with winter annual grain crops include thistles such as *Cirsium arvense* and poppies (for example, *Papaver rhoeas*). These are widely found around the Mediterranean. Some weeds prefer growing alongside summer annual crops (sunflower, maize and cotton), demonstrating a preference for aridity, for example *Amaranthus albus* and *Chenopodium album* (fat hen). These species belong to the Amaranthaceae and Chenopodiceae families, which are indicators of aridity in the pollen record and were predominant in the steppe communities of the Pleistocene cold stages in the Mediterranean. Perennial crops such as vines, oranges and orchards have typically been invaded by species such as pimpernel (*Anagallis arvensis*) and convolvulus (*Convolvulus arvensis*). Some crop invaders are tolerant of herbicides, which makes their control problematic, and those invaders which are bulbous, such as the Bermuda buttercup (*Oxalis pes-caprae*) mentioned in the introduction to this book, are favoured by ploughing as this helps to disseminate the bulblets. As well as cultivated fields, roadsides and ruins provide good sites for weeds or ruderal species. These are often nitrate-enriched environments. Since such environments are expanding with increased urbanisation around the Mediterranean, these plants are becoming more invasive.

Invasive plants have come to be associated with grazing lands as well as cultivated areas and so their advance has paralleled the spread of domestic livestock, particularly where there is overgrazing. The success of these particular invasive plants lies in the fact that they are less palatable, often being spiny (for example spiny broom, *Calicotome villosa*), poisonous (for example, black nightshade or *Solanum nigrum*) or just unpalatable (for example, *Cleome*

Plate 5.1 *Agave* (the century plant) and *Opuntia* (the prickly pear cactus) growing on the Likavittos Hill, Athens.

ambylocarpa). Many of these have spread on the drier steppes of northern Africa and the eastern Mediterranean.

Many of the plants associated with agriculture moved from east to west. Other species have come from other tropical or mediterranean-type ecosystems. Some of these are succulents such as the agave (*Agave americana* and *A. sisalana*), the prickly pear cactus (*Opuntia ficus-indica*) (Plate 5.1) – both from the Americas – and the hottentot fig (*Carpobrotus edulis*) from South Africa. The agave is sometimes known as the century plant, given its long period between flowering. It has large spiny leaves and tall (up to 6 m) inflorescences. In many parts of the Mediterranean it has been planted as impenetrable hedging. Apparently the spiny tip of the leaf, when still attached to leaf fibres, was used in the Algarve as a needle and thread in rural areas. Agaves have also been important sources of sisal (Mabberley and Placito, 1993). The prickly pear (*Opuntia* spp.) has also been widely used as a living hedge. It produces edible fruit and until the mid-nineteenth century the plant was an indirect source of cochineal, a carmine-red dye. The dye consists of the dried remains of scale insects of various species of the Coccidae family, which feed on the juices of the *Opuntia*, among other plants. The insects were harvested by scraping them into bags using long poles (to avoid the prickly spines on the cactus). Once in the bags the insects were boiled to extract the dye (Mabberley and Placito, 1993). The hottentot fig is a succulent, storing water in its narrow fleshy leaves so reducing the risk of dehydration. It has invaded many cliff and sand dune areas. It produces a fruit which is edible and the plant may have been introduced deliberately. It is invasive and thrives where native vegetation has been disturbed. It spreads vegetatively and from seed, and its mats overwhelm other plants, to the extent that it is now a danger to native flora.

Many tree species have been introduced either as ornamental specimens or for their use in forestry and some of these have become naturalised. Good examples are the acacias of the western Mediterranean, most of which were originally native in Australia and Tasmania but were planted as ornamentals or used for sand stabilisation – for example the silver wattle (*Acacia dealbata*) in France and Portugal, the blackwood acacia (*A. melanoxylon*) in Algeria, France and Portugal, and the golden wattle (*A. cyclops*) in Portugal (Le Floc'h, 1991). Several species of eucalyptus from Australia have also been widely planted for their fast growing timber and in places have become naturalised, for example *Eucalyptus camaldulensis* in the eastern Aegean and Anatolia in Turkey. Eucalyptus are fire-tolerant species and well adapted to mediterranean-type climates. Where burning is repetitive they can spread and impose their own ecological characteristics on the environment.

The significance of introductions is difficult to assess because it is far easier to recognise those that are successful than to have any records of those that disappear quickly. This is highlighted by one of the few detailed records tracking exotic species – the *Flora Juvenalis* referred to at the beginning of the book. These arrived with the wool trade near Montpellier in France between the seventeenth and nineteenth centuries. By 1859, 458 species had become established in the area but by 1950 only 6 still survived (Le Floc'h, 1991; Groves, 1991). The demise of the initially successful species came about because the technology by which they spread – the washing of seed-contaminated wool and its spreading to dry on nearby fields – ceased to be used. There may be similar occurrences of unknown exotics which arrived but have subsequently disappeared unnoticed. Estimates of the rate of introduction of new species in most regions with a mediterranean-type climate are between four and six per year (data quoted by Groves, 1991) but this would hide regional variations and also attempts to eradicate invasive introduced species, although only rarely is such eradication successful.

5.3 Historical development of Mediterranean animal populations

5.3.1 Mammals

Over 200 mammal species are estimated for the Mediterranean region – 197 actually recorded, plus an unknown number as yet unrecorded. Many of these also have a non-Mediterranean distribution and so the total number confined to the Mediterranean region is probably nearer to 155 (Cheylan, 1991). Species richness varies significantly between southern Europe, the Middle East

and North Africa as a result of biogeographical source area. The Middle East
is the richest of the regions, with 117 species, 23 of which are endemic. North
Africa has 91 species, while the Balkans, Italy and the Iberian Peninsula share
between 71 and 82 species. This distribution reflects the dispersal abilities and
migratory paths of the mammals, as shown by the fact that species affinity is
greatest between North Africa and the Middle East, while it is least between
North Africa and southern Europe.

Although the Strait of Gibraltar is only 14 km wide, it is an insuperable
barrier to non-flying mammals. The only occasion when mammalian exchange
was possible across the Strait and other areas of the Mediterranean Basin was
when sea levels were lower during the Miocene. Convergence of the African
plate with the Iberian Peninsula 6 million years ago blocked the outlet of a
proto-Mediterranean to the Atlantic Ocean. The Mediterranean Sea became
isolated from the world's ocean system and dried up as inflow from rivers
draining into the basin was exceeded by evaporation. This event is known as
the Messinian salinity crisis (Hsü et al., 1973). Only isolated brine lakes and
salt swamps survived in the deepest parts of the former sea. Since that time,
sea levels have not been low enough, even during Pleistocene cold-stage sea-
level regressions, for faunal exchange to take place.

In contrast to the western Mediterranean there has been a land connection
between Europe and the Middle East and North Africa for at least the last
12 million years. However, freedom of movement for mammals between these
regions has not always been easy given the changing environmental conditions
in the Middle East during Pleistocene climatic changes – between more desert-
like conditions as at present and more humid phases. The result of these
factors is that southern Europe's mammal species are derived mainly from the
Euro-Siberian sub-region of the Palaearctic (see Figure 5.2b) and so share an
affinity with northern and eastern Europe. In North Africa, most mammals
originate from sub-Saharan Africa and from the hot desert belt which stretches
from Morocco to India known as the Saharo-Sindian sub-region of the
Palaearctic. Given the location of the Middle East it is not surprising that its
mammals reflect origins in the Euro-Siberian sub-region, the Irano-Turanian
sub-region, the Saharo-Sindian sub-region, sub-Saharan Africa and Asia south
of the Himalayas.

About 25% of the Mediterranean's mammals are endemic, but endemism
is higher for island and peninsular mammal populations and in mountainous regions, such as those of North Africa. Long periods of unbroken isolation
and an originally depauperate fauna have promoted mammalian speciation.
An interesting example of this is the now extinct populations of pygmy or
dwarf elephants, hippopotami and deer found on Crete, Cyprus, Corsica and
Sardinia, and also giant rodents (Reyment, 1983). The early ancestors of these
animals are thought to have colonised islands by swimming and drifting with
longshore drift and ocean currents. Once isolated, evolutionary changes in
body size occurred probably in response to limited territory and food resources,
enhanced by genetic stress induced by a reduced gene pool (Schüle, 1993).
The island species became distinct from their mainland ancestral species. The

island elephant (*Elaphus falconeri*) was only a quarter the size (0.8–0.9 m tall) of its supposed ancestor, *E. namadicus* (4–5 m tall). As well as changes in stature, island elephants exploited new niches and habitats because of environmental stress resulting from a lack of traditional habitat and new environmental stimuli. Fossil bones on Crete show that dwarf elephants were capable of climbing rocky slopes, unlike their ancestors. In rodents gigantism may have resulted from an absence of predators. The colonisation events for these mammals were probably short-lived (Reyment, 1983) and, as there is no evidence for the mixed occurrence of bones of both ancestors and descendants, it is likely that there were no arrivals of ancestral forms after the new adaptations had become established. Parallel evolution patterns occurred on the different Mediterranean islands with large mammals becoming smaller and small mammals becoming larger, irrespective of the island involved. The continued isolation of these islands led to an unbalanced fauna. Once rejoined to the mainland during lower sea levels in the Pleistocene glacial maxima, invasions of mainland mammals which were more competitive in their food requirements resulted in the extinction of these dwarf and giant mammals and the reappearance of a balanced fauna. This disappearance pattern of fauna from an unbalanced population is a well-observed phenomenon in many different parts of the world (Schüle, 1993).

There is evidence that dwarf elephants and hippopotami survived into the Holocene, for example on Cyprus, and coexisted briefly with Neolithic people. More than 240 000 bones of the pygmy hippopotamus have been found at one Cypriot site (Vigne, 1992). At several sites bones of hippopotamus and elephant have been found with cut marks associated with the use of flint tools. Hence their demise is associated with early hunters (Simmons, 1988; Schüle, 1993). Worldwide this relationship is recognised as Pleistocene overkill – the hunting and eventual extinction of large mammals (Martin, 1984). Such definitive evidence for the coexistence of dwarf megafauna and *Homo sapiens* on Cyprus is not available from other islands, but Schüle suggests that in Lower to Middle Pleistocene times *Homo erectus* probably caused the extinction of elephants and hippopotami on Corsica/Sardinia.

Similar extinctions have also taken place in the mainland Mediterranean, especially during the Holocene and probably as the result of the degradation of habitat by cultivation. Areas with high endemism are especially vulnerable to species loss from human activity. This will be examined further in Chapter 9, but it is worth noting that between the end of the last glacial period and the end of the Roman occupation of many Mediterranean islands, there was considerable species turnover. As well as human-induced extinctions, there have also been many introductions – of domestic species such as sheep, goats, pigs, cattle, horses, dogs and cats; of game species such as red deer, hares, rabbits and the red fox; and small mammals arriving inadvertently by ship such as rats, mice and shrews. Mammal diversity is now two to five times as great as it would have been during the Pleistocene (Cheylan, 1991), although this increase has been accompanied by a decrease in genetic diversity since most of the endemic mammals have become extinct with human settlement.

The introduction of mammals has not been confined to islands. There are examples of several intentional introductions into the Mediterranean region (Cheylan, 1991). These would obviously include animals domesticated in the Near East approximately 10 000 years ago (see Section 8.2) but others too. The Moors are reputed to have introduced the genet (*Genetta genetta*) to Spain, possibly with the mongoose (*Herpestes ichneumon*) which is found there. In the Balkans, the mongoose probably invaded via the Black Sea. The Romans were responsible for bringing the fallow deer (*Cervus dama*) to Mediterranean Europe and the red deer (*C. elaphus*) to North Africa. More recent introductions are the coypu (*Myocastor coypus*, also known as the nutria) and the muskrat (*Ondatra zibethicus*) to Provence from the Americas in the 1940s. In some places, introduced animals such as sheep have escaped from captivity and become naturalised feral populations. Some mammals spread through the Mediterranean independently of people but are commonly found with people – namely the mouse (*Mus musculus*) and the black rat (*Rattus rattus*). The mouse entered the Levant from Asia sometime between 22 000 and 12 000 BP and the rat, surprisingly given its close association with people and their waste, is unknown in European deposits before 2200 to 2000 BP. In Egypt the species is known from 3500 BP (Cheylan, 1991).

5.3.2 Birds

Bird species richness is extremely high. Blondel (1991; Blondel and Aronson, 1995) gives figures of about 345 species of breeding birds, compared with 419 for the whole of Europe, an area more than three times the size of the Mediterranean region. The three main biogeographical regions from which these originate are the boreal forests of northern Europe and Asia, the southern and eastern semiarid steppes ranging from the Atlantic to central Asia (with which there is the closest affinity) and the Mediterranean itself, in which a small number of taxa has speciated. As with plants and other animals, climatic fluctuations and the dispersal ability of birds are important for the distributions of species. Blondel's work on biogeographical origin of bird species and ecological succession or disturbance shows that birds of Mediterranean origin are only important in early successional or immediate post-disturbance stages, those with structurally less complex vegetation communities. As time since disturbance progresses and vegetation matures, forest birds from boreal regions become widespread and dominant. This observation is also apparent in other old mature European forests and is interpreted as reflecting the importance of 'core' forest trees, those that have been widespread across Europe throughout the Pleistocene. As forest communities have developed so too have the bird communities. In the Mediterranean some areas of forest have survived through the Pleistocene climatic fluctuations, hence the success of birds of boreal origin. The survival of forest refugia and the continued existence of shrub vegetation have also been advocated as reasons for the low rates of avian endemism, about 13%. Given the good ability of many birds to disperse, there does not appear to have been the necessary geographical isolation between forests and steppe communities, in either the Mediterranean or other regions, for many examples of allopatric speciation to occur – speciation resulting from separate distributions of the same taxa. However, an interesting allopatric distribution is that of the partridge, which has also been successfully introduced to a number of Mediterranean islands, but each with its own revealing pattern.

Four different species of partridge (*Alectoris* spp.) exist in mainland Mediterranean – *A. barbara* in North Africa, *A. rufa* from Spain to Italy, *A. graeca* from Italy to Greece and *A. chukar* in Turkey and the Middle East. There is also a narrow belt of hybridisation between *A. rufa* and *A. graeca* (in Italy) and between *A. graeca* and *A. chukar* where these species come into contact (Blondel, 1991). There are no bone records which show the existence of any partridge on Mediterranean islands before human invasion about 10 000 to 8000 years ago, but there is evidence from historical records of more than one species of partridge being introduced to each of the larger islands (Blondel, 1991; Blondel and Aronson, 1995). However, today each of the larger islands has only one species, that of the mainland to which it is nearest – *A. barbara* in Sardinia, *A. rufa* in Corsica, *A. graeca* in Sicily and *A. chukar* in both Crete and Cyprus. The restricted island distribution is an example of competitive exclusion superimposed on the pattern of colonisation (Blondel, 1991).

5.3.3 Amphibians and reptiles

Rates of endemism of amphibians and reptiles in the Mediterranean region are higher than those of mammals and birds. Blondel and Aronson (1995) give figures of 56% and 62% for 62 and 179 species, respectively. It is probable that these high rates reflect the considerable length of time since their speciation before the Miocene-Pliocene, and the fact that they are sensitive to any changes in geographical barriers. Some reptiles and amphibians have been introduced, especially to islands: 7 of the 18 species on Corsica and as many as 13 of the 16 species in the Balearics were introduced. Amphibians and reptiles may be good indicators of environmental change (see Section 7.2.4).

5.3.4 Soil fauna

Far less is known about the soil fauna of mediterranean-type ecosystems than of the animal populations living on or above the ground surface. This gap in our knowledge needs filling because it is increasingly clear that some soil animals may be keystone species in ecosystem functioning – that is, they help to regulate the health of ecosystems. An example is the role of worms and their effects on the physical properties of soils, on soil biogeochemical cycles, in promoting the successful germination of vegetation and as a source of protein for predators (Blondel and Aronson, 1995). This particular example is examined further in Section 7.2.3, but it is increasingly clear that we need more information about soil animals and their distribution and diversity in the Mediterranean region. Some data are available which draw extensively on patterns found in other regions with a mediterranean-type climate (di Castri, 1973).

To some extent, soil fauna are less affected by the peculiarities of mediterranean-type climates than animals and plants above ground because the soil effectively moderates above-ground climates. Soil microclimates are more stable – conditions of soils at depth show similarities across climatic boundaries, except where there is extreme aridity or waterlogging (di Castri, 1973). In responding to seasonal climatic fluctuations most soil fauna adjust the timing of their life cycles or migrate through the soil layers (Andrés et al., 1999). However, soil animals may need adaptations to the nature of mediterranean-type vegetation, such as sclerophyllous leaves, which inhibit attack by soil arthropods (insects, spiders, woodlice, millipedes, centipedes etc.) and delay their decomposition. Nevertheless, there is a marked stratification of soil animals with respect to soil depth.

The greatest diversity of different soil animal groups is found in the litter and humus layers, where levels of organic matter are high. Mediterranean soil fauna is mainly composed of mites (Acari) and springtails (Collembola), with the former generally being more abundant (Andrés et al., 1999). In general the production of humus in a mediterranean-type climate is poor, with its amount varying with the density and characteristics of the vegetation, but it is also partly determined by the activities of the soil organisms themselves. Humus is also strongly modified by agricultural activities. These would therefore influence the soil faunal populations – their community diversity and population numbers. Deep soil layers tend to have fewer soil animals, in part because excess soil water at depth in many soils forces many so-called 'true soil animals' in mediterranean climates to seek refuge in arboreal microclimates. di Castri (1973) reports a very rich fauna under bark, moss, liverworts and lichens growing on trees. Such animals would more normally be found in the soil or in humus or litter in more humid environments, for example springtails (Collembola) and the primitive, blind insects of the class Protura. In a way this means that the term 'soil animal community' is a misleading simplification because there are strong seasonal patterns to the movement of arthropods, for example between the above- and below-ground environments (Andrés et al., 1999). The litter layer marks the boundary between the two. When moisture levels are high enough the litter is inhabited by arthropods more characteristic of deeper soil layers. But those which live in arboreal microclimates often retreat to the ground surface.

Because the relationship between soil fauna and vegetation type has been relatively little studied in the Mediterranean region it is difficult to assess whether land-use changes are altering distributions of soil fauna – without inventories of fauna in their 'natural' habitats how can we tell if faunal populations change as land use changes? An attempt to disentangle this question comes from the Languedoc region of France (David et al., 1999) focusing on millipedes (Diplopoda) and woodlice (Isopoda) living in the litter and upper soil layers and feeding on dead and decaying plant and animal tissue. They are locally very abundant in Mediterranean Europe and appear to have a more important role in recycling plant detritus than in more humid northern Europe. The study compared millipede and woodlouse diversity in quadrats along transects across a variety of vegetation communities – open ground sites including grassland, semi-open sites with mixed grassland and woodland, semi-open ground with Mediterranean shrub communities, and oak woodland. The results showed greatest diversity in the semi-open sites, with both a higher number of species and greater evenness than at open sites or under oak. This is a typical edge effect – the coexistence of species from both open and wooded sites in the ecotone between the two, with some species having no preference for any particular plant formation. These included *Cylindroiulus caeruleocinctus*, a species of millipede found in open habitats in moist temperate Europe but more common in Mediterranean Europe in wooded habitats. As a result this species is at risk because there is tendency in the region towards expansion of oak woodland. For conservation purposes some semi-open areas should be maintained within mosaics of habitat – grassland, shrubland and woodland. The conservation of saprophagous macroarthropods, like millipedes and wood-lice, is crucial in the decomposition of litter. Observations in the region suggest that the increase in oak cover is accompanied by decreases in the abundance of macroarthropods and accumulations of undecomposed leaf litter.

5.4 Adaptations of plants to Mediterranean climate

Mediterranean vegetation communities are dominated by three main components – sclerophyllous shrubs (drought tolerant), deciduous shrubs (drought evaders) and drought avoiding perennials and annuals. These are present in varying proportions depending on a number of factors such as substrate characteristics, fire and, in particular, scale of human activity. Although pollen analysis can sometimes help to reconstruct the 'natural' vegetation which existed prior to landscape degradation, the reconstructions cannot be complete. Some species, especially those which are insect-pollinated, are poor pollen producers and so are inevitably under-represented in the pollen record. Nevertheless, pollen records show that similar communities existed throughout much of the Mediterranean, with remnants still found today. The success of these communities lies in the adaptations of the vegetation and their life-forms and life histories to summer drought and winter rainfall. This climatic combination is not optimal for plant growth as high temperatures (optimal for photosynthesis) coincide with poor water availability (sub-optimal for transpiration). Hence there are relatively few tree and shrub species in mediterranean climates. Hemicryptophytes and therophytes are the most common life-forms (see Figure 5.1).

5.4.1 Sclerophyllous shrubs

Sclerophyllous plants are those with tough, leathery evergreen leaves. Although typical of many mediterranean-type plants they are found in other climatically stressful regimes where there is a need for plants to reduce transpiration and conserve water. Leaves are small with thick cuticles and often smaller leaves have a greater thickness, for example in the oaks, *Quercus coccifera* and *Q. ilex*

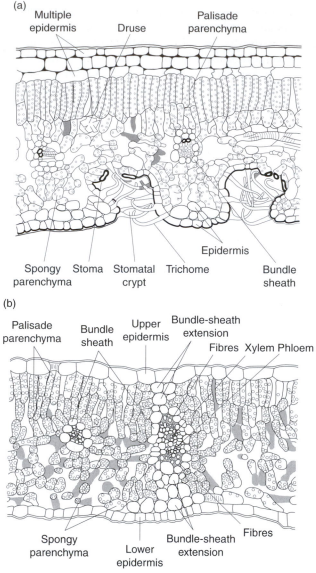

Figure 5.4 Transverse sections of (a) a sclerophyllous leaf of *Nerium oleander* and (b) a mesomorphic leaf of *Pyrus communis*. The sclerophyllous leaf has a thick cuticle, a multi-layered epidermis with thick-walled cells, tightly packed palisade layer cells and stomata lined with hairs. The mesomorphic leaf has a one-layered epidermis with thinner cell walls, a loosely packed palisade layer and no sunken stomata. Reproduced from Esau, K., 1965, *Plant Anatomy*, 2nd edition, reprinted by permission of John Wiley & Sons, Inc., New York.

(Kummerow, 1973). The stomata are sunken, generally into grooves or pits on the lower surface, and the openings are reduced by hairs or wax. These features can be seen in the cross-section of an oleander (*Nerium oleander*) leaf in Figure 5.4(a). Although the number of stomata may be high (up to

1000 per mm^2 compared with 100 per mm^2 for succulents) they are small (1–2 µm). The leaves also have strong, small, thick-walled cells that help maintain leaf shape as water is lost. These structural adaptations to survive water stress have been described as xeromorphism. The principal factors leading to a lower ratio of surface to volume have been identified as water deficiency, high light intensity and nitrogen deficiency (Kummerow, 1973).

In addition to leaf adaptation, sclerophyllous shrubs generally have smooth bark and upright branches, so that stemflow concentrates water to the base of the stems, where it infiltrates the soil and is available to the roots. Rooting depth is also crucial in determining water availability. Deep-rooting species are better able than those with shallower roots to penetrate groundwater reserves year round.

Adaptations in sclerophyllous plants are common in all mediterranean-type ecosystems – a feature long since recognised and debated. It is presumed that with very little commonality of species and distinct evolutionary histories, the evergreen sclerophyllous shrubs of mediterranean-type environments show convergent evolution. The selective forces are not just climatic, but also include fire and soil characteristics. These are discussed in more detail in Box 7.1.

5.4.2 Deciduous shrubs

Deciduous shrubs are drought evaders, producing leaves at the beginning of the wet season and shedding them with the arrival of summer drought. A comparison of the leaf structure of a pear (*Pyrus communis*) (Figure 5.4(b)) with that of the oleander (*Nerium oleander*) (Figure 5.4(a)) clearly shows large, thinner-walled cells, loosely packed and separated by air spaces in the *Pyrus*. A further characteristic of some deciduous shrubs is seasonal dimorphism. Leaf morphology occurs in two distinct forms – winter and summer leaves. The reduction in leaf area lowers water loss through transpiration by 50–76%. For *Sarcopoterium spinosum* and *Phlomis fruticosa*, Margaris (1977) found that the total amount of soluble sugar, chlorophyll, fat, nitrogen and free and bound amino acids was higher per leaf dry weight in winter leaves than in summer ones. Rates of photosynthesis and dark- and photo-respiration were thus higher in winter leaves. Also, the trigger for seasonal dimorphism may not be water stress itself, but seasonal changes in photoperiod, or the day length (Margaris, 1975). This is important as winter precipitation is irregular; if plants were to wait for winter rainfall to grow new larger leaves, more energy would be expended in leaf production than in making use of available water. By using day length as a trigger, leaf areas are optimal when winter rains occur.

5.4.3 Drought avoiders

Winter annuals are an important component of Mediterranean flora. These germinate in autumn, take advantage of winter rains and complete their life cycle, setting seed, before or at the start of the dry season. They thus avoid drought in the dormant state. Their importance is shown in Figure 5.1.

Other drought avoiders include geophytes, plants which survive the summer months below ground as bulbs, corms or rhizomes, with leaves and flowers being produced in winter and spring. Included in this group are many garden bulb plants familiar in more temperate climates, for example members of the iris (Iridaceae), lily (Liliaceae) and amaryllis (Amaryllidaceae) families. Such garden bulbs include tulips, daffodils and crocuses. Garden varieties have been bred from wild species but some have also been transplanted from the wild, threatening their survival. Conservation of these plants is discussed in Section 9.4.

5.4.4 Aromatic plants

The Mediterranean Basin is especially rich in aromatic plants, those which produce essential oils. This makes a walk through some plant communities a heady experience with a sweet perfume coming from many of the herbs familiar to anyone with an interest in cooking – thyme, mint, basil, parsley, fennel, sage, rosemary, lavender, coriander, rue, bay leaves, wormwood, fenugreek, sesame, saffron, licorice, onions, shallots, garlic and chives. Some 49% of the world's aromatic plants are found in the region (Ross and Sombrero, 1986) and the majority belong to the mint (Lamiaceae), carrot (Apiaceae) and daisy (Asteraceae) families. Why should the Mediterranean have such a rich aromatic flora? The answer is complex because the ecological role of the aromatic compounds in the leaves of these plants and to a lesser extent in roots and seeds is not fully understood. However, it is likely to lie in the stresses of a mediterranean environment.

The essential oils are produced in the leaves and the oil yields can be extracted by distillation. This is the basis for the perfume industry around the city of Grasse in Provence, France. It also allows for experiments on oil production in response to a range of environmental stimuli. For example, monthly yields of oil from a number of different Lamiaceae, including sage (*Salvia officinalis*), marjoram (*Origanum vulgare*), *Majorana hortensis*, balm (*Melissa officinalis*), basil (*Ocimum basilicum*) and thyme (*Thymus vulgaris*), peak during summer drought. This suggests a regulatory factor associated with leaf production in summer, but it is not clear whether this is aridity *per se* or whether aromatic oils are concentrated in summer leaves as a deterrent to predators, either insects or grazing animals (Ross and Sombrero, 1986). Many Mediterranean plants are lacking in nitrogen. They therefore have a low photosynthetic capacity – leaves take a long time to develop, at some 'cost' in terms of energy use to the plant. In addition, some Mediterranean soils are low in nitrogen and phosphorus, which also leads to slow growth rates, though with seasonal flushes in winter when moisture is available. A combination of these factors would suggest that to recover the costs of slow growth leaves need to be long-lived and the production of essential oils in the leaves is believed to promote this through the enhanced allocation of carbon into the biosynthesis of oils. As well as being a defensive mechanism against herbivores, the oils may be antimicrobial or antifungal, they may inhibit the germination of competitor seeds, or mimic insect pheromones as a way of attracting pollinators, they may cool the leaves through volatilisation of the oils and so reduce transpiration and, as the oils are highly inflammable, may also have a role in the spread and occurrence of natural fires. It is therefore likely that a variety of stress-related environmental factors has led to the evolution of aromatic plants. Indeed, some research shows that within just one taxon, thyme (*Thymus vulgaris*), there is considerable intraspecific variability in oil content across short distances between individual plants. This appears to be caused by slight changes in the growing environments of these individuals (Gouyon *et al.*, 1986).

5.5 Adaptations of plants to fire

Fire is common in mediterranean-type ecosystems. Although fire frequency, especially that of natural fires, is higher in other mediterranean-type ecosystems such as California and South Africa, some areas of the Mediterranean Basin burn regularly, generally as a result of human-induced burning. Thirty per cent of the French Mediterranean landscape is believed to have been burnt at least once between 1965 and 1990 (Blondel and Aronson, 1995). Where fires were once thought of as devastating to the environment, leading to loss of species diversity, degraded ecosystems and increased risk of soil erosion, it is now accepted that they are a natural part of ecosystem processes (see Section 7.4). Indeed, most species growing in regions where fire is common can be recognised as fire adapted, surviving fires and regenerating and reproducing after fires. However, it is difficult to be certain that adaptations are specific to fire, and not to other disturbances; certain adaptations may be more pronounced as a result of frequency or intensity of fire (Trabaud, 1987). It has also been suggested that as fire is integral to mediterranean-type ecosystems it may have acted to eliminate species which do not have adaptive traits, retaining those that do. Plants which respond to fire are known as pyrophytes.

Four categories of sexual or reproductive strategy for survival after fire can generally be identified – vegetative resprouting, increased stimulation of flowering, stimulation of seed germination and increased seed liberation and dispersal. Vegetative resprouting, either from below-ground biomass or from epicormic buds located on the trunks and branches of trees, is common in many taxa. Distinctions can be made between obligate and facultative resprouters (Naveh, 1974). Obligate resprouters rely solely on vegetative reproduction. They are summer-active, drought evaders such as the kermes oak (*Quercus coccifera*), the mastic or lentisc tree (*Pistacia lentiscus*) and the carob tree (*Ceratonia siliqua*) and include most sclerophyllous trees and shrubs of the Mediterranean Basin garrigue and maquis. An excellent example of a tree with fire-resistant, insulating, thick bark is the cork oak (*Q. suber*) with new shoots sprouting from the bark. Many species of eucalyptus also have fire-resistant bark and new shoots sprout from basal stumps (Plate 5.2). Facultative resprouters are able to regenerate both by resprouting and seed production. All are chamaephytes and drought evaders. Included in this category are the sage-leaved cistus (*Cistus salvifolius*), sage (*Salvia triloba*) and thyme (*Thymus capitata*),

Plate 5.2 Coppiced eucalyptus resprouting after a fire.

plants typical of garrigue and steppe communities (see Chapter 6). Members
of the Ericaceae family, such as the tree heath (*Erica arborea*) and the strawberry
tree (*Arbutus unedo* in the western Mediterranean and *A. andrachne* in the
east) are also facultative resprouters from root crowns rather than suckers.
Root regeneration in facultative resprouters usually occurs with the first rainy
season after a fire, while seeds are produced in the first post-fire summer
(Naveh, 1974).

  Taxa which require fire for seed germination are known as obligate seeders.
Typical species are the Aleppo pine (*Pinus halepensis*), the narrow-leaved cistus
(*Cistus monspeliensis*), the grey-leaved cistus (*C. albidus*) and rosemary
(*Rosmarinus officinalis*). Some have increased seed liberation and dispersal after
fires; in others germination of seeds is stimulated. Increased seed liberation
and dispersal is associated with taxa producing woody fruits. For example,
several species of pine have serotinous cones. These open and release seed only
when subjected to extreme fire temperatures. Aleppo pine (*Pinus halepensis*) in
the western Mediterranean readily replaces itself in areas which are frequently
burnt. So too do maritime pine (*P. pinaster*) and, in the eastern Mediterranean, *P. brutia*. This helps to explain the widespread dominance of pine
forests in some regions (Barbero *et al.*, 1998). In those plants where germination of seeds is stimulated by fire, seeds are released on burn or remain buried
in the soil awaiting fire. Fire can mechanically rupture seed coats, as in the
case of *Cistus* species. High temperatures may also inhibit fire-sensitive compounds in the soil which normally act to prevent germination. Fire can also
promote germination through the increased availability of light to the soil as

the fire eliminates the leaf canopy. Experiments have shown that germination of cistus, for example *C. monspeliensis* and *C. albidus*, is enhanced by a change in light quality (Roy and Soulié, 1992).

Sexual reproduction may be promoted by increased stimulation of flowering after fire. Many geophytes are included in this category, such as members of the amaryllis (Amaryllidaceae), iris (Iridaceae) and lily (Liliaceae) families. After flowering, seed production increases. Therophytes, or annuals, are also able to take advantage of post-fire site conditions of open ground and increased soil fertility, along with facultative resprouters and perennial grasses.

5.6 Summary

This chapter has examined the origins of the Mediterranean flora and fauna – especially the determining roles of climate and fire on their characteristics. Common associations of vegetation and animal communities are often found around the Mediterranean Basin and the relationships between these organisms and their physical environments are the subject of Chapter 6. The effect of fire, particularly fire frequency and intensity, on ecosystem structures and the dynamics of vegetation communities is discussed further in Chapter 7. The use of fire as a tool in landscape management is also considered in Chapter 9.

5.7 Further reading

Archbold, O.W. (1995) *Ecology of World Vegetation*, Chapman & Hall, London.

Chapter 6

Communities

The idea that characteristic groupings, or communities, of species are identifiable within climatic regions is one that is firmly established in Mediterranean ecology. This reflects the dominance of many French-speaking botanists in the study of Mediterranean vegetation from the 1930s onwards, for example Emberger (1930) and Braun-Blanquet (1932). Their work, mainly in southern France, followed Clementsian models of vegetation succession from bare ground through a herbaceous cover to maquis and then forest. Surveys were carried out in which the abundance, dominance and sociability of all individual species were recorded within field units or *relevés*. Each unit had uniform topography, soil and flora. This methodology, known as phytosociology, seeks to explain the ecology and dynamic characteristics and relationships that exist between different associations (see Box 6.1). Many French results form the basis of our present understanding of ecosystems around the Mediterranean and provide a background against which to assess the role of human impact.

Nevertheless, the notion that certain plants associate with each other as determined by environmental factors, which are largely climate-based, is over-simplistic. The concept of communities of plants and animals being in equilibrium with climate is giving way to non-equilibrium views in which 'nature' is seen as being more chaotic (Pahl-Wostl, 1995). Associations of plants and animals are not passive but respond to disturbances, whether local and relatively short-lived such as the occurrence of a wildfire, or longer term such as climatic change. Species that associated with each other in the past do not necessarily do so today, and vice versa. Species compositions fluctuate and may not be in equilibrium with prevailing climate. It is therefore difficult to accept uncritically the idea that it is possible to identify, with certainty, typical communities of Mediterranean plants. Yet most of the literature on the flora of the region has presupposed that it is. In some cases this occurs because it is unavoidable to use such terms as 'shorthand' when describing vegetation. It also helps non-specialists to make sense of the vegetated landscape, as long as it is remembered that communities are not static but dynamic. It is in this sense that vegetation communities are described in this chapter. However, before this can be done we need to address the issue of terminology; this bedevils many attempts to study the literature on Mediterranean vegetation.

Box 6.1   Phytosociology

Sclerophyllous shrubland, together with oak woodlands, has been the focus of considerable research. Not surprisingly given its geographical location, much of the early research was carried out by phytogeographers from Montpellier, France, and much of their methodological approach followed that of the Zurich–Montpellier school of vegetation classification, or phytosociology (Braun-Blanquet, 1932). This approach is less commonly used in English-language biogeographical literature than in France. Indeed, acceptance of this approach to the study of vegetation communities is not widespread in other areas with mediterranean-type climates, such as California, and the methodology of phytosociology has been much criticised. However, for the Mediterranean Basin phytosociology has been widely used and its nomenclature underpins much of the literature.

Problematically for those people unfamiliar with its use, the terminology is complex, but in essence phytosociology seeks to identify and record all plants present within an area, based on field surveys of floristic *relevés*, or field units, which have uniform topography, soils and flora. Within each field unit the abundance, dominance and sociability of individual species are recorded.

Phytosociology also includes a study of the ecology and dynamic characteristics and relationships that exist between different associations. Designated associations (and their sub-associations and variants) are then placed within a hierarchy of alliances, orders and classes. In order to indicate the hierarchical level of analysis, suffixes are added to the generic names of the plants after which the communities are named. For example, in the western Mediterranean, the class Ononido-Rosmarinetea (named after the genera *Ononis* and *Rosmarinus*) is subdivided into a number of different orders, such as Rosmarinetalia, Anthyllidetalia (named after *Anthyllis*) and Phlomidetalia purpureae (named after *Phlomis purpurae*) (Quézel, 1981). Part of the difficulty in basing vegetation analysis on phytosociology in the Mediterranean lies in the fact that the number of plant associations runs into hundreds, which may differ one from another on the basis of the presence or absence of a single characteristic species or the predominance of certain species (Tomaselli, 1981a). Despite its complexities, Quézel and Barbero (1982) argue in favour of phytosociology, in that identification of plant associations helps in ecological zoning, especially for landscape planning and conservation.

6.1   Mediterranean vegetation communities

6.1.1   A plethora of terms, or how to be confusing – Part I

Perhaps the most commonly described vegetation communities of the Mediterranean are the evergreen sclerophyllous shrublands dominated by evergreen oaks (such as *Quercus ilex*), olives (*Olea*), carobs (*Ceratonia*), *Pistacia*, *Cistus*

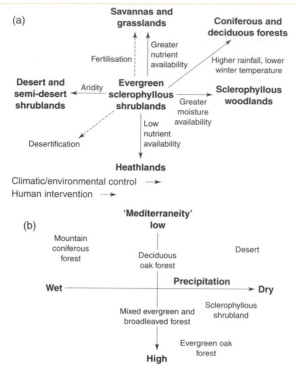

Figure 6.1 (a) Environmental gradients and Mediterranean vegetation communities. After di Castri (1981). (b) The six main potential vegetation communities found in the Mediterranean region. After Mazzoleni *et al.* (1992).

and many labiate herbaceous species. These communities are described in vast detail by di Castri *et al.* (1981) and are recognised as comprising one of the ecosystems of the world. The evergreen sclerophyllous shrublands are set at the centre of different mediterranean communities which occur with change in environmental gradients such as moisture availability, nutrient availability and cool temperatures (see Figure 6.1(a)). In other words, climate is regarded as the main determinant of vegetation type, together with substrate, which determines nutrient availability. At the extremes of these gradients, but still regarded as mediterranean-type ecosystems, are the following communities: savannas and grasslands, coniferous and deciduous forests, heathlands, sclerophyllous woodlands, and desert and semi-desert shrublands.

Climatic criteria have also been used in attempts to define communities, based on the early presumption that vegetation communities are in equilibrium with environmental factors, in particular the summer period of effective physiological drought. As a result calculations of potential evapotranspiration and several derivative indices are used. A classic approach was that of Emberger (1930; Nahal, 1981) mentioned in Box 2.1. This gives six different bioclimatic types based on aridity, each further subdivided according to eight thermal variants. Corresponding to these are vegetation zones such as a subhumid mediterranean vegetation zone. However, these are further distinguished by an altitudinal component to give, for example, alti-Mediterranean, oro-Mediterranean, cryo-Mediterranean and Alpine-Mediterranean according to degree of elevation and cold winter temperatures (see, for example, Le Houérou, 1990). The result is complexity and confusion of terms. Yet these are widely used in much of the literature on Mediterranean vegetation, especially that derived from French-based research. However, as far as possible these terms are avoided in this book.

Identification of such bioclimatic vegetation types is, of course, crucially dependent on which climatic parameters are used and how they are defined. Different parameters will yield different types and distribution patterns. To overcome these difficulties, multivariate statistical analysis of rainfall and temperature data has been used to map Mediterranean climate variations and relate these to vegetation types. Using cluster analysis and principal components analysis, Mazzoleni et al. (1992) identified seven main potential vegetation types according to two principal components, one a precipitation gradient, the other a measure of 'Mediterraneity' (Figure 6.1(b)). 'Mediterraneity' is 'the climatic variability sufficient to maintain a winter precipitation maximum and good spring regeneration, yet with summer temperatures which risk summer drought stress and winter temperatures which risk cold (frost) stress' (Faulkner and Hill, 1997, p. 253). Thus it is based on seasonal differences in precipitation and temperature regime, with high scores representing a high degree of variability. Of the seven main vegetation types, six have a major distribution in the Mediterranean, as shown in Figure 6.2: sclerophyllous shrubland, deciduous oak forest, evergreen oak forest, mixed evergreen and broadleaved forest, mountain coniferous forest, and desert.

6.1.2 A plethora of terms, or how to be confusing – Part II

Confusion about different vegetation communities also arises because different countries of the Mediterranean use local names, such as maquis, jaral, matorral and phrygana. Fortunately, Tomaselli (1981a) codified the variety of names for shrubland, as shown in Table 6.1. He favoured the term matorral, which is characterised by shrubs whose aerial parts are not differentiated into trunks and branches, whose leaves are sclerophyllous, and whose growth habit may be upright or prostrate. Matorral is subdivided into high matorral (>2.0 m), middle matorral (0.6–2.0 m) and low matorral (<0.6 m). Ground cover may be dense, discontinuous or scattered, and trees may be present. Low and dense matorral may consist of thorny xerophytes. All of these varieties are reflected in the different names for shrubland across the basin. Most English-language-based literature refers to sclerophyllous shrubland as maquis, or as garrigue for the lower growing, often more degraded shrubland (Plate 6.1), and these terms are used in this chapter except where there is explicit reference to other people's work and their terminology.

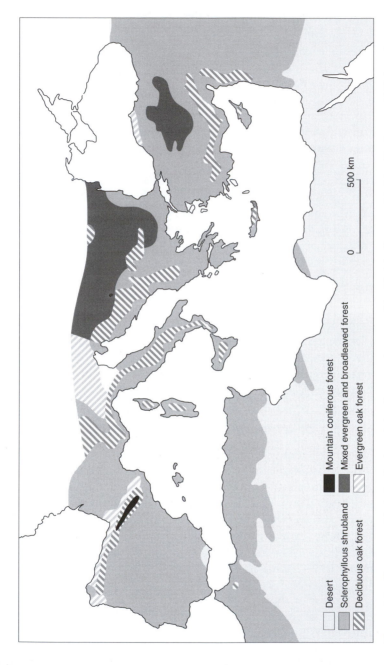

Figure 6.2 The geographical distribution of the six vegetation types identified in Figure 6.1(b). After Faulkner and Hill (1997).

Mountain coniferous forest

Mixed evergreen and broadleaved forest

Evergreen oak forest

Desert

Sclerophyllous shrubland

Deciduous oak forest

500 km

0

Table 6.1   Terms used to describe Mediterranean sclerophyllous shrubland (Tomaselli, 1981a; Margaris, 1981)

	High matorral (not usually subdivided on the basis of substrata)	Middle matorral	Low matorral
France	*Maquis* (xerophilous, sclerophyllous and evergreen low trees, generally very dense and impenetrable)	*Garrigue* (sometimes only with reference to calcareous substrata)	Less well defined, but sometimes *garrigue* (or *landes* on siliceous substrata)
Spain	*Matorral denso, espinal*	*Matorral claro, jaral* (on siliceous substrata)	Less well defined, but *tomillar* is used
Italy	*Macchia alta*	*Macchia bassa*	*Garriga (gariga)*
Greece			*Phrygana*
Eastern Mediterranean			*Batha*
Other MTEs California South Africa Australia	*Chaparral* *Fynbos* *Mallee*	*Scrub*	*Coastal sage*

Plate 6.1   Garrigue/steppe vegetation. Note the areas of bare ground between the cushion-shaped, low growing vegetation.

6.2 Plant and animal communities

Plant and animal communities do not, of course, live in isolation from each other. Birds, insects, small mammals etc. all have their preferred habitat and within these make specific use of plant matter, whether as food or shelter.

Table 6.2   Regional variation in numbers of species (in brackets) in some characteristic genera of Mediterranean sclerophyllous shrubland

Western Mediterranean (e.g. Morocco)	Circum-Mediterranean	Eastern Mediterranean (e.g. Turkey)
Genista (22)	*Anagyris*	*Sideritis* (about 20)
Helianthemum (22)	*Artemisia*	*Hypericum* (about 15)
Teucrium (14)	*Calicotome*	*Salvia* (about 15)
Cistus (12)	*Ceratonia*	*Satureja* (about 12)
Cytisus (12)	*Fumana*	*Phlomis* (about 10)
Thymus (10)	*Jasminum*	*Rhamnus* (about 7)
Thymelaea (9)	*Juniperus*	*Ebenus* (about 5)
Halimium (8)	*Laurus*	*Daphne* (about 3)
Erica (7)	*Lonicera*	*Globularia* (3)
Adenocarpus (6)	*Myrtus*	*Astraglus* (impossible to determine)
Asparagus (6)	*Olea*	
Coronilla (5)	*Osyris*	
Lavandula (at least 5)	*Phagnalon*	
Ulex (4)	*Phillyrea*	
	Pinus	
	Pistacia	
	Quercus	
	Rhamnus	
	Rosmarinus	
	Ruscus	
	Spartium	
	Staehelina	
	Viburnum	
	Ziziphus	

Source: adapted from Quézel (1981).

Animals and plants interact as part of ecosystems, as discussed in Chapter 7. Unfortunately, research on Mediterranean animal populations has taken place in a degree of isolation from that of plant ecology and there are few syntheses of the animal ecology of the region. However, some of the more common associations of plant and animal communities are examined in the following sections on the characteristic vegetation communities. In general birds are excluded from this immediate discussion but are considered separately in Section 6.3.

6.2.1 Sclerophyllous shrubland or maquis

Sclerophyllous shrubland is the most widespread vegetation type found around the Mediterranean Basin, usually where precipitation levels range from 350 mm to 1500 mm per annum (Faulkner and Hill, 1997). It is absent from the coastal regions of Libya and Egypt. Typical plants include olive (*Olea*), carob

Plate 6.2 *Cistus ladanifer* (gum cistus). This species occurs in two forms, with or without five purple spots at the base of the petals.

(*Ceratonia*), *Pistacia*, *Cistus* and many labiate herbaceous species. Some genera are found throughout the region but the widespread distribution of maquis belies significant variations in floristic composition between the western and eastern regions of the Basin (Quézel, 1981); there are more species of *Erica* and *Cistus* in France and Morocco, for example, but more species of *Salvia* and *Phlomis* in Greece and Turkey (see Table 6.2). As well as regional variations in type of shrubland, it is possible to distinguish between maquis and garrigue on calcareous and non-calcareous substrates, for example the various species of *Erica* or *Cistus* found on non-calcareous soils in the western Mediterranean.

Four species of oak and two of pine are particularly common in sclerophyllous shrublands – holm oak (*Quercus ilex*), kermes oak (*Q. coccifera*), cork oak (*Q. suber*), *Q. calliprinos*, Aleppo pine (*Pinus halepensis*) and Calabrian pine (*P. brutia*). Their distributions may be regional rather than circum-Mediterranean; for example, holm oak does not grow in Turkey and Calabrian pine is not found in the western Mediterranean. Other conifers also grow in sclerophyllous shrublands, such as junipers (*Juniperus oxycedrus, J. phoenicea*) and alerce (*Tetraclinis articulata*), which is restricted to the southern part of the Iberian Peninsula and northern Africa.

Additional important trees are olives (*Olea europaea*) and *Pistacia* (the lentisc or mastic tree, *P. lentiscus*, in the eastern Mediterranean, and the terebinth or turpentine tree, *P. terebinthus*). These may all be found in communities with or without the carob (*Ceratonia*). Across the Mediterranean there are also the strawberry trees (*Arbutus unedo* and *A. andrachne*, the latter found only in the eastern Mediterranean) and species of *Cistus* (Plate 6.2), *Phillyrea* and *Laurus*. There is some indication of an increase in the arborescent elements of sclerophyllous shrubland in the eastern Mediterranean (Quézel, 1981).

A variety of medium to large mammals is found in maquis communities. These include rabbits (*Oryctolagus cuniculus*) and red deer (*Cervus elaphus*). In the Iberian Peninsula maquis is important habitat for remaining populations of the Iberian lynx (*Lynx pardinus*), whose conservation is discussed in Section 9.3.5a. Small-mammal diversity is relatively low (Tucker and Evans, 1997). By contrast there is a rich and abundant reptile fauna. Typical species include the globally threatened Hermann's tortoise (*Testudo hermanni*), spur-thighed tortoise (*T. graeca*) and marginated tortoise (*T. marginata*). Lizards include the agama (*Agama stellio*) and Lilford's wall lizard (*Podarcis lilfordi*), while snakes include the large whip snake (*Coluber jugularis*) and the western whip snake (*C. viridflavus*).

A fairly characteristic assemblage of birds is found in maquis, including warblers (*Sylvia* spp.), buntings (*Emberiza* spp.) and partridges (*Alectoris* spp.). Some birds are locally common such as the black francolin (*Francolinus francolinus*) in Turkey and Cyprus; others are more widespread such as shrikes (*Lanius* spp.) and finches (*Carduelis* spp.) (Tucker and Evans, 1997). More open ground, such as garrigue, is important for diurnal and nocturnal raptors attracted by high densities of prey. Such birds of prey include the short-toed eagle (*Circaetus gallicus*), the long-legged buzzard (*Buteo rufinus*), golden eagle (*Aquila chrysaetos*), kestrel (*Falco tinnunculus*), eagle owl (*Bubo bubo*), Scops owl (*Otus scops*), little owl (*Athene noctua*) and Bonelli's eagle (*Hieraaëtus fasciatus*). There is also a range of insectivorous birds such as the nightjars (*Caprimulgus europaeus* and *C. ruficollis*), the bee-eater (*Merops apiaster*) and the roller (*Coracias garrulus*). In rocky and dry grass habitats characteristic species include the short-toed lark (*Calandrella brachydactyla*), crested lark (*Galerida cristata*), thekla lark (*G. theklae*), tawny pipit (*Anthus campestris*), black-eared wheatear (*Oenanthe hispanica*) and several species of partridge (*Alectoris* spp.) (Tucker and Evans, 1997).

6.2.1a Maquis – climax community or not?

The presence of trees in some areas of maquis and garrigue has introduced the debate about the position of maquis in a successional sequence. In some instances maquis has been interpreted as a climax vegetation community; elsewhere it is regarded as a stage in succession, either progressive or retrogressive. Many of these ideas are exemplified in the work of Tomaselli (1981b) and Barbero *et al.* (1990). However, the application of succession theory to mediterranean ecosystems is increasingly viewed as inappropriate. This is based on the argument that mediterranean vegetation development is not progressive and unidirectional and therefore successional; instead, frequent disturbances result in perturbation-dependent communities (see Section 7.4). The argument is compelling but not fully accepted and much of the literature still reflects the influence of Clementsian successional ideas.

Where maquis and garrigue are considered as climax communities the two characteristic species are olive (*Olea oleaster*) and carob (*Ceratonia siliqua*), often found in association with *Pistacia lentiscus*, Mediterranean buckthorn (*Rhamnus alaternus*) and *Phillyrea angustifolia*, although this association depends

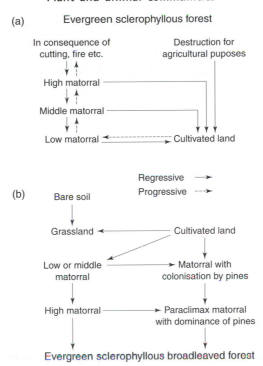

(a)   Evergreen sclerophyllous forest

In consequence of   Destruction for
cutting, fire etc.   agricultural puposes

High matorral

Middle matorral

Low matorral ←------------→ Cultivated land

Regressive →
(b)   Bare soil   Progressive -->

Grassland ←   Cultivated land

Low or middle   → Matorral with
matorral   colonisation by pines

High matorral ——→ Paraclimax matorral
                   with dominance of pines

Evergreen sclerophyllous broadleaved forest

Figure 6.3 (a) Theoretical regressive and progressive succession from or to forest. After Tomaselli (1981b). (b) Theoretical progressive succession with pines which tend to replace (paraclimax) forests of hard broadleaved evergreen species. After Tomaselli (1981b).

on geographical distribution. In the southern Iberian Peninsula, the dwarf palm (*Chamaerops humilis*) often grows in association with *Olea* and *Ceratonia*; on the northern coast of Crete, the Theophrastus palm (*Phoenix theophrasti*) is typical. From Greece eastwards, the strawberry tree (*Arbutus andrachne*) is an important associate. In southern Spain and Portugal, climax maquis may also be dominated by the tree heath (*Erica arborea*). Climax maquis or matorral extends along the coasts of northwest Spain, southern France, central-southern Italy, southern Greece, western and southern Turkey, Israel, the Mediterranean islands and parts of North Africa (Tomaselli, 1981b).

In the view of succession-based analysts most maquis is not climactic, but represents a stage in a progressive succession to forest or a retrogression from forest. Through fire, excessive grazing and felling of climax trees, high or dense matorral degrades. Matorral may also become established where cultivation is abandoned. Figure 6.3(a) illustrates the generalised theoretical successional sequence (Tomaselli, 1981b), which can be compared with Figure 6.3(b), where matorral is colonised by pines and eventually progresses to evergreen sclerophyllous broadleaved forest, often dominated by holm oak (*Quercus ilex*). The presence of pine trees is especially important in progressive successions, and pine communities are considered in more detail below (Section 6.2.3).

At the wetter and cooler winter margins of its distribution, sclerophyllous shrubland merges into forest cover. This may be the cool-temperate deciduous forests further north in Europe or coniferous forests at higher altitudes, both more typical of non-mediterranean-type climates, but the transition may be to broadleaved evergreen forests where holm oak (*Quercus ilex*) is almost the only canopy tree. This forest type is considered next.

6.2.2 Evergreen oak forest

Holm oak (*Quercus ilex*) may be present in shrubland or maquis communities and it has leaves similar to those of the evergreen shrub oaks *Q. coccifera* and *Q. calliprinos*. However, at the wetter and cooler peripheries of maquis (in Emberger's cold semiarid and temperate humid Mediterranean bioclimates) holm oak may form continuous cover (Terradas, 1999). This oak is a circum-Mediterranean species, especially abundant in the western Mediterranean, but absent from those regions where summer drought is too sustained. Thus in the drier parts of the basin, such as North Africa, southern Spain and southern Greece, it is mainly found at higher elevations. Holm oak forests are particularly extensive in Spain, Morocco, France, Italy, Corsica and Sicily. Much of the research on the ecology of holm oak forest has been carried out in France and Spain (Rodà *et al.*, 1999).

Holm oak may form mixed forests with Aleppo pine (*Pinus halepensis*) or stone or umbrella pine (*P. pinea*) but it tends to be the dominant tree species in closed forests. In mixed forests the pines are frequently taller than the oaks (Terradas, 1999). Other trees and shrubs may also be present such as *Arbutus unedo*, *Phillyrea latifolia*, *Rhamnus alaternus* and *Pistacia lentiscus*. Climbing plants may be important, such as ivy (*Hedera* spp.), honeysuckle (*Lonicera* spp.) and clematis (*Clematis* spp.). The grass cover is rarely very extensive.

Until at least the 1970s, closed-canopy holm oak forests were thought of as climax communities and capable of self-regeneration. It is now clear that the tree has been closely favoured by people. Pollen analysis suggests that in southern France it became dominant only after 2000 BP and spread with the increase in human activity and the rise of agriculture in Gallo-Roman times (Terradas, 1999). Today few forests are undisturbed; most bear evidence of cultivation and coppicing (Rodà *et al.*, 1999).

Ecologists are interested as to whether holm oak forests are capable of self-regeneration. Observations show that when forests are well established, there are few oak saplings growing in the understorey, which suggests a problem in the recruitment of younger trees. At the same time there appear to be large numbers of seedlings. The question is, therefore, what happens to the seedlings to prevent them from growing into saplings?

Holm oaks grow slowly, forming dense canopies. Under reduced light levels and associated higher soil moisture, germination of acorns is stimulated and survival of seedlings is good – experiments indicate that high light levels do not increase growth rates in seedlings, presumably because the reduced leaf area ratio offsets any increase in photosynthesis (Retana *et al.*, 1999). This makes the seedlings shade-tolerant and they can survive for years without significant growth beneath the canopy. Some 55% of a seedling's weight is in root biomass in shady environments – a considerable investment in non-photosynthetic structures, leading to high maintenance costs and so small net carbon gain and slow growth rates. However, once a disturbance occurs seedling survival is seriously affected and few are able to grow on to the sapling stage.

The main disturbance factor is fire, which eliminates most seedlings, except those which are very large (Retana *et al.*, 1999). Acorns are also unlikely to survive fire because of desiccation. As a result regeneration of holm oak after fire is dependent on resprouting of older trees rather than germination of seeds. Acorn production and establishment of new seedlings are subsequently delayed until the resprouts reach reproductive age. Regeneration by resprouting of stumps is further aided by their already-existing root systems; resprout growth rates are high in the first year after a disturbance (Terradas, 1999). If seedlings are prevented from growing into saplings and the number of resprouting individuals is high enough to complete regeneration, then does it matter that this is the principal method? At present the answer is debatable. Resprouting does not introduce new genotypes into a forest (Retana *et al.*, 1999). In addition, resprouting stumps gradually become senescent (aged), which might lead to low production and eventual degradation of holm oak forests. However, many holm oak forests in France possibly date from the Middle Ages and as yet show no sign of decline.

Although the ability of holm oaks to regenerate from seedlings is rare under closed canopies because of the absence of old seedlings, seedling survival rates increase in mixed Aleppo pine and oak stands as light levels increase slightly. As a result holm oak saplings become more abundant. In turn this leads to a gradual replacement of pine by oak and the long-term dominance of oak. Young pines are less able to establish themselves in a shaded understorey provided by an oak canopy (Retana *et al.*, 1999). The progressive recruitment of young oaks beneath Aleppo pine lends some support to traditional ideas of vegetation succession from pine to oak as outlined in Figure 6.3(b).

Holm oak forests have been managed since early historical times. On poor soils, as in central and western Spain and in Portugal, holm oak woodlands have been managed as *dehesas* (or *montados* in Portugal). These are savanna-type ecosystems with trees growing in a generally non-treed landscape. Their management over the centuries was traditionally for their firewood, for their acorns for livestock, as shelter for livestock and as pasture for sheep and cattle. Today they have a conservation value because of their high biodiversity. *Dehesa* systems are described further in Sections 8.5 and 9.3.4. Closed holm oak forests have also been managed since at least Neolithic times, generally as coppice plots (Terradas, 1999). Charcoal and tannin production used to be important but have now declined. Today their main use is for firewood.

Some large herbivores are found in broadleaved evergreen oak forests, such as red deer (*Cervus elaphus*), roe deer (*Capreolus capreolus*) and wild boar (*Sus scrofa*). Large carnivores such as the brown bear (*Ursus arctos*), the wolf (*Canis lupus*) and the two species of lynx (*Lynx lynx* and *L. pardinus*) are generally rare and found only in a few locations (Tucker and Evans, 1997). The bird fauna of broadleaved evergreen oak forests is richer than that of Mediterranean coniferous forests. There are important breeding populations of raptors such as the honey buzzard (*Pernis apivorus*), short-toed eagle (*Circaetus gallicus*), lesser spotted eagle (*Aquila pomarina*) and the globally threatened Spanish imperial eagle (*A. adalberti*). Kites (*Milvus milvus, M. migrans*), flycatchers (*Muscicapa striata* and *Ficedula semitorquata*) and warblers (the olivaceous warbler *Hippolais pallida*, the olive-tree warbler *H. olivetorum*, Bonelli's warbler *Phylloscopus bonelli*) are all found in evergreen and deciduous oak forests (Tucker and Evans, 1997).

6.2.3 Coniferous forests

Coniferous forest vegetation is found widely around the Mediterranean Basin, especially in mountainous areas, and pines form a major component. Cedar forests are found in the Atlas Mountains (*Cedrus atlantica*), in Cyprus (*C. brevifolia*) and in Turkey and Lebanon (*C. libani*). Companion species include maples (*Acer monspessulanum* and *A. obtusatum*), holly (*Ilex aquifolium*) and yew (*Taxus baccata*). Firs occupy varying areas in high mountain regions; *Abies pinsapo* in Spain, *A. nebrodensis* in Sicily and Calabria, *A. cilicica* in Turkey and *A. cephalonica* in Greece (Le Houérou, 1981). In addition there are junipers (*Juniperus oxycedrus, J. phoenicea*) and alerce (*Tetraclinis articulata*), which is restricted to the southern part of the Iberian Peninsula and northern Africa.

The large mammal diversity of coniferous forests includes the wolf (*Canis lupus*), wild cat (*Felis sylvestris*) and pine marten (*Martes martes*). In some high-altitude forests chamoix (*Rupicapra rupicapra*) and wild sheep (*Ovis ammon*) may be found (Tucker and Evans, 1997). Birds of Mediterranean coniferous forests are similar to those of non-Mediterranean coniferous regions, being dominated by a number of ubiquitous species such as woodpigeons (*Columba palumbus*), turtle doves (*Streptopelia turtur*), firecrests (*Regulus ignicapillus*) and green woodpeckers (*Picus viridis*). There are a few species restricted to Mediterranean coniferous forests (in comparison with broadleaved forest) – the Corsican nuthatch (*Sitta whiteheadi*), Krüper's nuthatch (*S. krüperi*) and the crossbill (*Loxia curvirostra*) (Tucker and Evans, 1997).

6.2.3a Pine forests

It is estimated that while pines cover only 5% of the total Mediterranean area, they comprise about 25% of forest cover and up to 75% in North Africa and Anatolia (Barbero *et al.*, 1998). About ten different pine species are found in the region with four described as 'Mediterranean shore and island pines' (Klaus, 1989), that is those with a true Mediterranean distribution – Aleppo pine (*Pinus halepensis*), Calabrian pine (*P. brutia*), Canary Island pine (*P. canariensis*) and stone or umbrella pine (*P. pinea*). The original distributions of the other species were in the central European and Asiatic mainlands from which they spread to the Mediterranean – dwarf mountain pine (*P. mugo*), black pine (*P. nigra*), maritime pine (*P. pinaster*), Scots pine (*P. sylvestris*), *P. uncinata* and *P. heldreichii*.

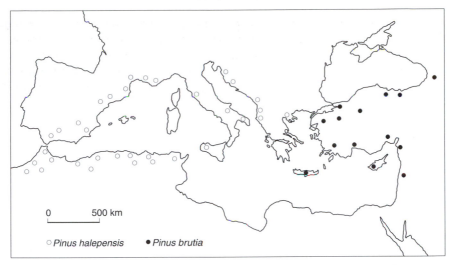

Figure 6.4 The present distribution of *Pinus halepensis* and *P. brutia* in the Mediterranean region. After Biger and Liphschitz (1991).

There are marked differences in the distributions of the pine species reflecting variations in bioclimate, altitude and substrate. Of particular interest in terms of their morphology is the distribution of the two most common pine species, Aleppo pine (*P. halepensis*) and Calabrian pine (*P. brutia*), which together cover about 6.8 million hectares (Barbero *et al.*, 1998). Genetically and ecologically they are very similar yet have vicariant distributions. *P. halepensis* is found in the western Mediterranean and *P. brutia* to the east (see Figure 6.4). They coexist in two small areas of Greece, in southeastern Anatolia and in the Lebanon and Israel. Here they form natural hybrids (Barbero *et al.*, 1998). Climate is the likely determinant of their distributions, for geological, archaeological and historical evidence suggests an eastern origin for both (Biger and Liphschitz, 1991). *P. halepensis* spread west in response to its climatic needs and climatic change. It requires a more humid and warmer climate than *P. brutia*, not being able to survive temperatures of −10 °C, which *P. brutia* can. However, *P. brutia* may well be a relatively recent introduction to Israel and surrounding areas for its absence from Israeli archaeological records suggests its absence from the natural vegetation landscape during the late Holocene, until recent planting.

In terms of aerial coverage (about 3.5 million hectares), black pine (*Pinus nigra*) is the third most important pine species, found especially in mountainous regions and on a variety of substrates. Its maximum extent is in the Balkans and Anatolia, where there are numerous subspecies and varieties (Barbero *et al.*, 1998). At low and medium altitudes umbrella pine (*Pinus pinea*) and maritime pine (*P. pinaster*) are relatively common. *P. pinea* is widely planted and so its natural range is difficult to establish but it is found throughout the region, except in North Africa, especially on sandstones and sandy soils. *P. pinaster* generally prefers acid soils, and tolerates dolomites. It is restricted to the western Mediterranean. Pines growing in high mountain regions include *P. heldreichii* (southern Italy, northwest Balkans and eastern Greece), *P. uncinata* (western Alps, Pyrenees, Sierra de Javalambre) and *P. mugo* (scattered though the southern Alps and Balkans). The Scots pine (*P. sylvestris*) is not a true Mediterranean pine but has naturalised, generally on marls, limestones and dolomites, with well-established populations in Spain and southern France.

Pinus halepensis has been identified as a post-disturbance coloniser (Lepart and Debussche, 1991). In southern France it often grows on newly abandoned agricultural land. Its success owes much to its ability to produce cones with viable seeds from about 12 years of age. Seeds are produced abundantly, with seed densities of up to 25 seeds/m^2/yr at the edge of a mature woodland. Although seed dispersal is greatest over short distances, some dispersal will occur over long distances because the seeds are winged. Rapid germination of seeds is favoured by the high light levels present in recently abandoned fields. Invasion of *P. halepensis* is greatest near to existing stands with a ready seed source, but saplings growing around isolated trees are also foci for new invasions. Frequent disturbance is needed for its continued presence. Once a herbaceous understorey develops, germination is severely limited. Consequently *P. halepensis* is a transient species, unable to replace itself. It is usually replaced by oaks such as *Quercus ilex* and *Q. pubescens*.

Although pines are common in forest communities, there has been debate as to whether they form climax communities. Early ideas (1930s) rejected the notion of pine climax communities in the region, but a few authors regard some pine forests as remnants of previous pine climax forests, or as representing paraclimaxes, also known as plagioclimaxes or disclimaxes (Barbero *et al.*, 1998). These are the result of earlier periods of sustained human activity modifying pre-existing pine climax communities or the replacement by pine of broadleaved evergreen forest (Tomaselli, 1981b). Each pine paraclimax is associated with a broadleaved tree which represents the true climax community. For example, *Pinus halepensis* occurs with *Quercus ilex* and *Q. rotundifolia*, while in the eastern Mediterranean *P. brutia* occurs with *Q. caliprinos*. Elsewhere, *P. pinaster* is associated with *Q. suber* and *P. sylvestris* with *Q. pubescens*.

The extent of pine forests has changed considerably in the last fifty years or so. In the European Mediterranean, pine forests have been increasing in area, both as a result of changing land use – afforestation – and the ability of pines to invade abandoned or burned land and become 'weeds' – reforestation. In contrast, the area of pine forest in the Maghreb has declined through clearing of forests for cultivation, felling of timber for construction and charcoal production, overgrazing and frequent fires (Barbero *et al.*, 1990, 1998).

6.2.4 Savanna grassland, steppe and desert shrubland

By contrast to other areas with a mediterranean-type climate such as southern California and Australia, savanna grassland around the Mediterranean Basin is generally regarded as anthropogenic in origin, through clearing of trees and shrubs to create grazing pasture. This is especially so in upland areas, but may also be true of areas which were originally high matorral (Tomaselli, 1981b).

Perennial grasses coexist with perennial shrubs and winter annuals to form a mosaic of vegetation communities, which are successful in persisting from year to year despite heavy grazing, fire, nutrient-poor soils and intense summer drought. Such mosaics may be dynamic rather than static in their regeneration. This means that regeneration is two-dimensional – where one sub-community will be replaced by another, with lateral movement and replacement – as opposed to one-dimensional, in which a sub-community replaces itself on its site through the stages of youth, maturity and senescence or old age (Clark *et al.*, 1998). One-dimensional regeneration is regarded as destabilising, whereas two-dimensional regeneration tends to confer stability as the components of the mosaic replace each other, so reducing differences (for example in soil conditions) between the different stages of regeneration (youthful, mature, senescent).

The two-dimensional hypothesis is illustrated by an example of a community of perennial shrubs and annuals from the Rambla Honda ephemeral river catchment in southern Spain (Clark *et al.*, 1998; Puigdefábregas *et al.*, 1998). On the upper slopes of the valley, the perennial species is a tussock grass (*Stipa tenacissima*), a slow-growing drought and high-temperature endurer. It mainly regenerates vegetatively and demonstrates slow lateral replacement. It only rarely establishes seedlings in the inter-tussock areas. Instead the inter-tussocks are invaded by annuals, until the tussock grass has spread vegetatively into these areas to establish new tussocks, while the old have moved into the senescent stage. The new inter-tussock areas, where the previous tussocks have died, are then invaded again by annuals. Annuals are thus important because they are present at all stages of the tussock communities' development, albeit at different densities and biomass. The growth of the annuals aids community resilience and helps to prevent soil erosion following drought, fire or overgrazing. If the regeneration were one-dimensional, areas of bare ground would persist for longer, with the potential for greater surface runoff and sediment yield.

In arid areas of the Mediterranean, grasslands decline and steppe vegetation communities are recognisable. Tomaselli (1981b) cites three typical localities: Zaragoza in Spain, central Anatolia and Tunisia. In all three, steppe communities are regarded as degraded matorral and grasses such as *Stipa* and *Brachypodium* are common together with wormwood (*Artemisia*). Steppe communities, indicating aridity, were more extensive around the Mediterranean at the height of the Last Glacial Maximum in northern Europe, some 18 000 years ago.

In the most arid zones of the Basin, such as the Near East and North Africa, desert shrublands are found. For example in Jordan, there are open communities of juniper (*Juniperus phoenicea*), with a sub-stratum of wormwood (*Artemisia herba-alba*) and members of the pea family such as *Ononis matrix* or *Lygo raetam*. These too are regarded as degradational (Tomaselli, 1981b). The more arid areas of the Maghreb and the Middle East are home to small mammals, such as the gerbils (family Cricetidae).

6.2.5 Heathlands

On poor siliceous soils where nutrient availability is low Mediterranean heathlands are found. In France these are known as *landes*, although some-times this name is restricted to the sandy soils of temperate France. The ecology of Mediterranean heathlands is poorly understood and their extent was barely mentioned by Specht (1979) in his review of global heathland ecology. However, they are recognised in phytosociological studies by their typical vegetation associations, including the genera *Erica*, *Cistus*, *Quercus* and *Genista*. In acid soils phosphorus is less available than in neutral to alkaline soils so, in general, productivity is low in nutrient-poor soils. Experiments in France have shown that when nutrient-poor soils are fertilised productivity increases, demonstrating that the soils and vegetation communities are nutrient-limited (di Castri, 1981).

In the Mediterranean region heathlands generally occur as isolated islands surrounded by more calcareous substrates. Extensive studies of these commu-nities have begun in Spain, in particular the heathlands of the Gibraltar Strait region (Ojeda *et al.*, 1995). Preliminary results from floristic surveys show the importance of heathland in terms of the number of endemic species present. This could make them special targets for conservation. However, while rates of endemism may be high, species richness appears to be low due to the nutrient-poor status of the acid, sandy soils.

6.2.6 Wetlands

Even though they are not solely typical of mediterranean-climate regions, wetland communities are found in most Mediterranean countries and their importance to indigenous wildlife and migrating wildfowl is recognised; many are protected by the Ramsar Convention (see Chapter 9). Notable wetlands are found in the marismas (annually flooded marshlands) of the Coto Doñana bordering the estuary of the Guadalquivir river in Spain, the Camargue of the Rhône delta in France, Amvrakikós Gulf in Greece, Lakes Bardawīl and Burullus in Egypt and Garaet el Ichkeul and Sebkhet Kelbia in Tunisia. Some of these wetlands, such as the Camargue, have long been exploited for salt production but today their existence is threatened further by encroachment and reclama-tion for agriculture and urban land, by pollution from industrial, agricultural and domestic sources, by tourism and by falling water tables.

Wetland environments differ widely according to the depth of the water column, the duration of flooding and water conductivity. This produces a continuum between dry, more freshwater marshland and saline coastal marshes, more frequently inundated. In drier marshes, where there is only a brief flooding period during wet winters, plant communities are more terrestrial in character. In springtime annual plant species may be important, taking advantage of available moisture – for example *Cotula coronopifolia* (buttonweed, a native of southern Africa, but naturalised through the region), *Juncus* spp. (rushes), *Spergularia* spp. (sea-spurry), *Polypogon* (annual beard-grass) and *Hordeum* (wild barley). As soils dry out through the summer, so perennial species become important such as members of the Chenopodiaceae (goosefoot) family – *Sarcocornia* spp. and *Arthrocnemum* spp. (glassworts). Wetter marshes are controlled by water column depth. In shallow waters flooding may be limited to a few months in winter when root and bacteria demands for oxygen in soils are at minimum. In contrast deeper waters (more than about 10 cm) are dominated by flood-tolerant perennials. Species belonging to the genera *Typha* (reedmace), *Juncus* (rush), *Scirpus*, *Carex* (sedge), *Eleocharis* (spike-rush), *Ranunculus*, *Baldellia* and *Elatine* (waterwort) may be among the main components.

Vegetation communities also vary with water conductivity or salinity. Freshwater communities may include *Callitriche* spp. (starwort), *Potamogeton* spp. (pondweed), *Myriophyllum* spp. (milfoil), *Ranunculus* spp. and *Juncus* (rush). More saline communities may be dominated by species of Chenopodiaceae such as *Sueda maritima* (annual sea blight), *Salsola soda* (saltwort) and *Beta macrocarpa* (beet) together with *Puccinellia fasciculata* (tufted salt-marsh grass) and *Frankenia* spp. (sea heath).

Wetlands support abundant bird populations and many are staging posts for migrants, such as the wetlands of the western Algarve used by birds following a migratory route off the west coast of Portugal. These include sea birds and raptors, such as the osprey (*Pandion haliaetus*), marsh harrier (*Circus aeruginosus*), Montagu's harrier (*C. pygargus*) and hen harrier (*C. cyaneus*), which all remain for some time or overwinter in the western Algarve (Pullan, 1988). In the marshes of the Coto Doñana of Spain there are important populations, in terms of size, of greylag geese (*Anser anser*, 75 000), teal (*Anas crecca*, 50 000), wigeon (*A. penelope*, 40 000), shoveler (*A. clypeata*, 25 000), mallard (*A. platyrhynchos*, 20 000), pintail (*A. acuta*, 11 000) and tufted duck (*Aythya fuligula*, 11 000) (Garcia Novo and Merino, 1993). In addition there are some interesting species such as flamingo (*Phoenicopterus ruber*), spoonbill (*Platalea leucorodia*), cattle egret (*Bubulcus ibis*) and little egret (*Egretta garzetta*), herons (*Ardea* spp. and *Nycticorax* spp.) and storks (*Ciconia ciconia* and *C. nigra*). Increasingly wetlands are threatened by reclamation, therefore so too are their animal populations – birds, small mammals, reptiles etc. This is examined further in Section 9.3.4.

6.3 Bird communities

While some of the characteristic bird species of different vegetation communities have been included in the previous sections it is also interesting to examine the ways in which bird communities respond to changes in the structural arrangement of vegetation. This is one of the main components in habitat selection by birds. Several characteristics can be used to describe this arrangement, for example the average maximum height of the habitat, the number of vegetation strata present, the density of foliage cover, the horizontal and vertical diversity of the vegetation, and the complexity of the vegetation. In mediterranean-type ecosystems these characteristics distinguish between open grassland, low garrigue perhaps dominated by the kermes oak (*Quercus coccifera*), different types of maquis, for example with tree heath (*Erica arborea*) or arbutus (*Arbutus unedo*) as dominants, and holm oak (*Q. ilex*) woodland (Blondel, 1981).

Censuses of vegetation and bird populations along gradients from open to closed habitat in southern France and Corsica have shown that as complexity increases in the direction of *Q. ilex* woodland so too does the number of breeding bird species (Blondel, 1981). In each habitat different birds can be expected. The increase in number of species results from the appearance of new resources as habitats increase in height and foliage cover – the number of realised niches is proportionately linked to habitat complexity, in other words more niches are available in the more complex woodland habitats. Some species occupy quite wide niches that overlap with those of other species, for example the jay (*Garrulus glandarius*), the cuckoo (*Cuculus canorus*), the red-legged partridge (*Alectoris rufa*) and the sparrowhawk (*Accipiter nisus*). Others have much narrower niches such as Montagu's harrier (*Circus pygargus*), the little bustard (*Otis tetrax*) and the pin-tailed sandgrouse (*Pterocles alchata*). This is significant for the conservation of these species. Those which have a narrow niche in threatened habitat are more likely to become endangered. Montagu's harrier, the little bustard and the pin-tailed sandgrouse are all found only in open grassland, whereas those with a wider niche are found in a variety of maquis and woodland communities. Conservation of bird populations is very closely linked to conservation and management of their appropriate habitat.

In trying to determine the importance of habitat for bird populations, some studies are sufficiently detailed to show the seasonal dynamics of bird communities according to vegetation type. In a holm oak (*Quercus ilex*) forest in northeastern Spain a marked difference in bird communities occurs through the year, with maximum densities and species richness in the summer months and minimum in the winter (López-Iborra and Gil-Delgado, 1999). 'Constant species', those resident at least 50% of the time, dominate the forest, even though total species numbers increase in summer and decrease in winter. This conforms with other findings in the Iberian Peninsula, which show relatively few long-distance migrants in the breeding populations of evergreen forests in the Peninsula compared with deciduous forests. The likely explanation comes from the fact that in evergreen forests just a small fraction of leaves is renewed each year, and only young leaves may be eaten by the caterpillars which are the main food for breeding birds. By contrast, in deciduous forests there is complete renewal of leaves each spring and a plentiful supply of food for insects and therefore birds. Indirectly, therefore, climatic and environmental factors control species diversity through abundance or shortage of food caused by lack of or sustained stress, in this case summer drought.

6.4 Mammals

Some mammal species have already been described in their characteristic habitats but many have not survived until present times. Some 200 species have been identified for the Mediterranean region since classical times, though not all are native, some having been introduced. Several large mammals have been eradicated by human activity in the past 3000 years. The lion (*Panthera leo*) was present in Europe (in Thrace and Macedonia) until the first century AD. It survived in Turkey until 1870, in Syria until 1896 and in Morocco until 1922. The cheetah (*Acinonyx jubatus*) also survived in the Maghreb until 1930. The African or wild ass (*Equus asinus*) survived until Roman times in North Africa, while the onager (*E. onager*) became extinct in Syria in the early twentieth century. The African elephant (*Loxodonta africana*) is known to have lived in Morocco until the third century AD and possibly as late as the seventh century. The Asian elephant (*Elaphus maximus*) became extinct in the Orontes Valley (Turkey/Syria border) in the eighth century BC. The Bubal hartebeest (*Alcephalus busephalus*) became extinct in North Africa some time between 1930 and 1950 (Cheylan, 1991).

Populations of large mammals throughout the Mediterranean region remain severely threatened by habitat fragmentation, and indeed many are now restricted to a few isolated regions, such as the brown bear (*Ursus arctos*), the serval (*Felis serval*), the caracal (*Felis caracal*), the Iberian lynx (*Lynx pardinus*), the panther (*Panthera pardus*), the hyena (*Hyaena hyaena*), the Barbary sheep (*Ammotragus lervia*), the fallow deer (*Cervus dama*) and Cuvier's gazelle (*Gazella cuvieri*).

Families and species of smaller mammals are much more widespread, including those of bats, mice, shrews, squirrels, hares and the rabbit. For the most part, they are less threatened by habitat fragmentation. Indeed the rabbit (*Oryctolagus cuniculus*) is only indigenous in the Iberian Peninsula, but has been introduced and spread widely around the world. Some small mammals, such as the gerbils (family Cricetidae), have a more restricted Mediterranean range, being found most commonly in the more arid areas of the Maghreb and the Middle East. There is also one non-human primate in the region, the Barbary macaque (*Macaca sylvanus*). During the Pleistocene it was widespread throughout Europe and North Africa (Fa, 1986). Its disappearance from Europe is generally attributed to the Pleistocene glaciations, though anthropogenic activity may have been significant as well. In North Africa the retreat of its range from as far east as Egypt and Libya to Algeria and Morocco, where it is now found, is certainly due to human-induced habitat fragmentation. It now survives in small, disjunct populations but in diverse habitats, including maquis, evergreen oak forests and coniferous forests.

Some mammals are protected by virtue of their desirability as hunting targets. This is the case with wild boar (*Sus scrofa*), especially in the Iberian Peninsula. Its protection in the *dehesa* and *montados* woodlands also ensures the protection of its habitat and other species found with it (Abbott *et al.*, 1992). In North Africa it is the only fairly abundant large mammal because consumption of its meat is prohibited by religious laws (Cheylan, 1991).

The plant and animal communities of the Mediterranean Basin are the resources of the Mediterranean world on which civilisations have been built. The way in which land has traditionally been used, and still is used in places, is the subject of Chapter 8. However, before that Chapter 7 examines the ecosystems of the region.

6.5 Further reading

di Castri, F., Goodall, D.W. and Specht, R. (eds) (1981) *Ecosystems of the World*, vol. 11: *Mediterranean-type Shrublands*, Elsevier, Amsterdam.
Tucker, G.M. and Evans, M.I. (1997) *Habitats for Birds in Europe*, Birdlife International, Cambridge.

Chapter 7

Ecosystems

'Turning to soils that allow sprouting and bear crops a good distinction is made that refers to their production of grain or trees: fat soil is a better producer of grain, leaner soil of trees. For . . . grain (and indeed all annuals) get the food that is near the surface. But this food must not be scanty or dry out easily, as in lean soils. Trees on the other hand, because of their great and powerful roots, can also draw food from far below. But in fat soils this food is far too abundant, and whereas it produces a fine foliage and good height in trees, it produces no fruit, since the tree does not fully concoct the food. But in leaner soil the food turns out to be of just the right amount for producing both, and the trees master it and so are able to produce fruit. Extremely fat soil is good for no plant, drying up more than is wanted . . . The best-tempered soil is evidently the best, being in general of open texture, not cold, and containing moisture. In this way it is not only easily penetrated by the roots, but also feeds the plant well, these being the aims of land reclamation, tillage and manuring.'

Theophrastus (1976, II, 4.2 and 4.3)

Theophrastus, in fourth-century BC Greece, produced one of the earliest ecological texts, comparing cultivated plants with those of the wild. He gave detailed accounts of the best way to achieve maximum yields in wheat, how to derive cultivated olive trees from wild stock and where best to plant vines. He was familiar with the pollination of the fig – an elaborate symbiosis between the plant and its pollinator, the fig wasp (*Blastophaga psenes*). His text is a detailed work of research and although he never used the term it is clear from all he wrote, as in the quotation above, that he understood the concept of the ecosystem – the relationship between plants and animals and their environment. At the time he lived many of the modern crops of the Mediterranean world were already being grown and Theophrastus was aiming to produce a manual for agriculture – how best to manipulate wild species for the benefit of people. Cultivation and grazing had already created the many semi-natural vegetation communities which exist today – sclerophyllous shrubland, evergreen oak forests, savanna and grassland – some dating back to Neolithic times, like the *dehesa* woodlands of Spain.

The concept of the ecosystem is a convenient way to study the interaction of plants, animals and their environments. The term, like that of 'community', is generally recognised as being a mental construct rather than a readily identifiable object – that is, an ecosystem only exists if we think it does. Nevertheless ecosystem studies and attempts to understand ecosystems have become central to environmental concerns. From the 1950s there has been a widespread belief that understanding ecosystem processes promises a way to manage complex natural and semi-natural systems and understand how they respond if disturbance happens. Although we now recognise this to be idealistic, and there has been much discussion about the concept (see for example Golley, 1993), it remains a way of trying to understand the dynamics of plant and animal communities and assess the consequences of our actions, recognising what Trudgill (2000) calls the 'consciousness of interconnectedness'. Ecosystem studies therefore focus on ecosystem processes. The general principles of these are outlined in the next section; readers familiar with this material can skip it and go straight to Section 7.1.2, on ecosystem functioning in regions with a mediterranean-type climate.

7.1 Ecosystem functioning

7.1.1 General principles

Communities of organisms interact with their physical, non-living world. Through photosynthesis carbon dioxide is built up by plants (also known as autotrophs) into organic compounds. The total amount of energy produced per unit of time by photosynthesis is gross primary production, but a proportion of this is consumed by the respiration of plants. The difference between gross primary production and energy lost in respiration is net primary production. This is the energy available to organisms that eat plants, both living and dead plant tissue, and is then available to those animals that eat other animals. Thus 'autotrophs produce organic material, the consumers eat it, and the decomposers and detritivores clean up the mess (excrement and organic remains)' (Huggett, 1998, p. 142).

Ecosystem processes are the flows of energy and the biogeochemical cycles that allow ecosystems to function – that is, the cycling within the system of gases (such as carbon dioxide, oxygen and nitrogen), non-gaseous compounds (such as potassium, phosphorus, calcium and sulphur) and water. Such cycles are termed nutrient cycles if the minerals concerned are essential to life. The feeding relationships, or the flow of energy through ecosystems, produce food webs or food chains. At each stage of the food chain, or trophic level, energy is lost. Ultimately only a minute proportion (typically less than 1%) of the energy fixed in gross primary production is available at the top of the food chain. This means that there are generally fewer consumers than producers, and far fewer consumers at the top of a food chain or web than consumers nearer to the base of that chain or web. Some consumers are numerous by virtue of being very small, for example the many varieties of soil fauna.

As well as studying the storage and exchange of the terrestrial elements and compounds that make up ecosystems, ecologists and ecogeographers are interested in population and community processes within ecosystems. These are the interactions among organisms and the adaptive response of individual species to their environments, therefore determining which species form communities and assemblages according to a given set of environmental conditions. In other words, ecologists and ecogeographers ask the question 'what are the factors which help to explain why plants and animals live where they do?'. The answers relate to such variables as climate, natural disturbances (wildfires, tectonic activity) and anthropogenic factors (human-induced fires, land-use changes, species introductions, ecosystem management). Of increasing concern are the consequences of human activity (many of which have been unforeseen) such as induced changes in vegetation, soil erosion, invasions of exotic species and extinctions of native species, and changes in biodiversity. These are the topics of the remaining chapters. Land use and environmental and conservation concerns are the focus of Chapters 8 and 9, but first some key ideas about ecosystem functioning need to be explored in the context of the Mediterranean.

7.1.2 Ecosystem functioning in regions with a mediterranean-type climate

Net primary production (NPP) in mediterranean-type ecosystems is low by comparison with other terrestrial ecosystems. Forests have the greatest NPP, with an estimated 11 000 g/m^2/yr, and within this category tropical forests are the most important, with an NPP of 2300 g/m^2/yr (Ajtay *et al.*, 1979). NPP in chaparral, maquis and brushland is estimated at just 800 g/m^2/yr. Savanna, temperate grasslands, swamps and marshes, temperate woodland, cultivated land, and bogs and unexploited peatlands are estimated to have higher figures for NPP. Nevertheless mediterranean-type regions have been, and are, important in terms of ecosystem functioning given their role as major locations for human activity and because of their high levels of biodiversity (see Section 7.2). Key factors that have been considered in mediterranean-type ecosystems are the role of nutrients (Kruger *et al.*, 1983), fire (Moreno and Oechel, 1994), biodiversity (Davis and Richardson, 1995), global change (Moreno and Oechel, 1995) and landscape disturbance (Rundel *et al.*, 1998).

7.1.2a The role of fire

Fire is common in mediterranean-type ecosystems. Evidence for human-induced fires in the Mediterranean landscape goes back at least as far as the Neolithic (Naveh, 1975) and fire is still a major determinant of the landscape. Relatively few of today's fires are natural, for example from lightning strikes; most are caused by people, either intentionally or through negligence or out-of-control pasture burning, and a proportion have an unknown cause. Figure 7.1 illustrates this pattern for Spain where, as in the rest of the Mediterranean region, fire frequencies are increasing. Since the early 1960s the number

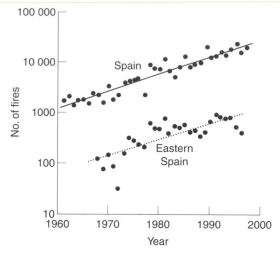

Figure 7.1   The number of fires for the whole of Spain and for the eastern part (Valencia), 1960 to the late 1990s. Adapted from Pausas and Vallejo (1999).

of fires and the surface area burnt in the European Mediterranean has increased exponentially (Pausas and Vallejo, 1999). This increase is in spite of fire suppression through fear of fire. Instead recent socio-economic changes in European Mediterranean countries have been advocated for increased fire numbers – in particular rural depopulation, increased agricultural mechanisation, decreased wood gathering and grazing pressure (and therefore declining burning of pasture to promote new growth) and increased urbanisation of rural areas. These changes have led to net abandonment of land and the recovery of vegetation (see for example Barbero *et al.*, 1990). As a result there is greater build up of litter as fuel, so that when fires do occur their impact is more severe, both in terms of widespread coverage and of intensity. This pattern is not dissimilar to other areas of the world where fires are a natural part of ecosystems but their occurrence has been suppressed. However, this pattern is not repeated in the southern Mediterranean where traditional land-use practices have continued, albeit with increased deforestation, but it is predicted that such trends will occur there in the future (Pausas and Vallejo, 1999).

The role of fire lies in both ecosystem functioning and landscape maintenance. In terms of ecosystem functioning the effects on soils are especially important (Pausas and Vallejo, 1999). Burning leads to direct loss of nutrients by volatilisation of carbon, nitrogen and sulphur, and by ash convection in the smoke column. Burning also temporarily clears the plant cover. This exposes topsoils to heating and to wind and water erosion. Fires have been shown to change soil texture, with a decrease in clay content and therefore increase in coarser particles (Vega and Diaz-Fierros, 1987). At the same time, fire creates water repellent layers at a few centimetres depth in the topsoil. This reduces infiltration and increases surface runoff and erosion. Soil losses have been measured as being higher on burned land than on unburned control plots, though the rates of soil loss vary with type of vegetation cover and the species present (see Section 4.5).

Following fires, nutrients may also be leached from the soil before vegetation is able to recover. This means that nutrient losses may be greater in the post-fire period than during the fire, especially when fires occur in summer and are followed by autumn rainstorms, which constitute a high erosion risk. Pausas and Vallejo (1999) quote research from Spain which suggests that it may be more than 1.5 years before vegetation cover spreads sufficiently to protect soils. Where fires take place in nutrient-poor ecosystems, there is a serious risk of depletion of soil fertility with decreased availability of nutrients such as inorganic phosphorus, and also reduced accumulation of carbon limited by losses of nitrogen during the fire. Reduced fertility could lead to a reduction in biomass production. Although little evidence has been found of this in the long-term where there are repeated fires, there is some indication that in stands of *Quercus coccifera* (the evergreen kermes oak) less biomass accumulates after 3.5 years in areas where fires are frequent (up to three fires in sixteen years) than in areas affected by only one fire (figures cited by Pausas and Vallejo, 1999).

In opposition to nutrient losses through burning, the incorporation of ash into soil raises fertility temporarily (Kutiel and Naveh, 1987). There is also a gradual increase in soil biological activity, although it may take more than twenty years for soil biological activity to return to pre-fire levels. This slow rate of recovery is associated with the disappearance of decomposers in the soil as a result of the fire. Some decomposers are very sensitive to fire. Blondel and Aronson (1995) quote figures of over 80% for loss of some decomposers during fires in the Pyrenees. Where fire frequencies are shorter than twenty years, some soil communities may not achieve a recovery equilibrium. However, not all decomposers are so sensitive to fire. Those that feed directly on dead plant material may, in fact, recover more quickly with the availability of burnt plant matter. These differing responses mean that it is difficult to generalise about the effects of fire on ecosystem processes and wrong to extrapolate from one set of community members to others. This can be seen in the consequences of fire on the greater white-toothed shrew (*Crocidura russula*) and the wood mouse (*Apodemus sylvaticus*). In shrublands of southern France, populations of shrews are displaced or destroyed completely by fire, while those of the wood mice survive and increase in density in the immediate post-fire period (Blondel and Aronson, 1995). This is the result of a short-term increase in the primary productivity of herbaceous plants and hence a food source for the mice. The wood mouse population increase is likely to stem from immigration of individuals from unburnt areas as well as from reproduction within the burnt areas.

The effects of fire on vegetation can be studied in terms of the response of individual species (reviewed in Section 5.5) and the response of communities. Fire has traditionally been regarded as a check to ecological succession and therefore as a factor promoting regression. More recently it has become accepted that the frequency of fire in mediterranean-type ecosystems makes it crucial to the functioning of these ecosystems and therefore their maintenance. While this is discussed in detail below in Section 7.3.2, at this point it is important to state that suppression of fires appears to reduce the richness of vegetation species and, in turn, the associated animal communities. Reports from southern France indicate that the turnover of bird species after a fire parallels that of the changes in vegetation communities (Blondel and Aronson, 1995). Bird species richness is high in the immediate post-fire phase and then declines as the vegetation becomes more uniform. It may increase again as more trees grow and introduce a more complex architecture for the birds.

If suppression of fires reduces species diversity by removing the 'rejuvenating' properties of fire then reduction in diversity may itself alter the way in which ecosystems function. The role of diversity is the subject of the next section.

7.2 Biodiversity – its role in ecosystem functioning

The role of biodiversity lies in its relationship with ecosystem functioning. We need to know how energy, nutrients and water flow within ecosystems. In addition, we need to identify the relevant organisms and the number and strength of interactions within populations and among species that ensure the uptake, utilisation and throughflow of solar energy and other resources (Blondel and Aronson, 1995). In other words, how do different organisms regulate ecosystems and what is their spatial and temporal distribution?

> 'There are various interpretations of what is meant by "biodiversity", and its constant use and misuse in the media has induced a negative reaction to the term in some sections of the scientific community, leading to its rejection as a serious scientific topic. The popularization of declining biodiversity has unfortunately put it in the category of a "flavor-of-the-month" issue when in fact it is a serious and difficult problem that deserves long-term scientific consideration.'
>
> (Walker, 1992, p. 19)

Explanations of patterns of biological diversity at a range of temporal and spatial scales are fundamental to ecogeography. It is therefore vital to consider what is meant by the term biological diversity, in other words biodiversity. In its most common usage, it refers to species richness, or the total number of species within a given area. However, 'species richness *per se* is an incomplete measure of biodiversity, but . . . it is nevertheless a necessary and useful benchmark of analysis of ecological complexity' (Davis and Rutherford, 1995, p. 346). In other words, we need to go beyond merely stating how many species there are in a given locality to look at further measures of diversity.

7.2.1 Biodiversity – hierarchical considerations

A useful starting point for analysing diversity is to look at 'the amount of genetic information which occurs in nested sets of living systems, from local

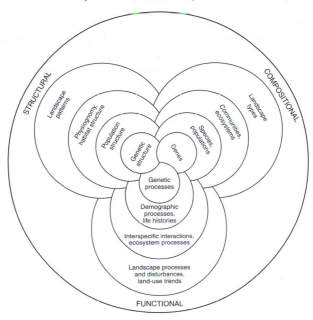

Figure 7.2 Compositional, structural and functional biodiversity, shown as interconnected spheres, each encompassing multiple levels of organisation. *Source:* Noss (1990).

populations to whole ecosystems, and sets of ecosystems (landscapes)' (Blondel and Aronson, 1995, p. 44). This points to the hierarchical nature of the subject, which is illustrated in Figure 7.2, showing the different levels at which diversity can be organised – genetic, population-species, community-ecosystem and regional landscape – and three attributes of ecosystems – composition, structure and function (Noss, 1990). Composition relates to the identity and variety of elements at each level and would include the traditional definition of diversity as species richness. Structure is the physical organisation or the pattern of a system, in other words habitat when measured at the community scale, or patches/mosaics of habitat at the landscape scale. Function involves evolutionary and ecological processes such as gene flow, disturbance and nutrient cycling. Implicit within this hierarchical approach is the notion that higher levels incorporate and constrain the behaviour of lower levels: 'if a big ball (e.g., the biosphere) rolls downhill, the little balls inside it will roll downhill' (Noss, 1990, p. 357). In other words, the effects of environmental change will permeate to all levels. The approach does not, however, advocate only the assessment of higher levels, arguing that '"big picture" research on global phenomena is complemented by intensive studies of life histories of organisms in local environments' (Noss, 1990, p. 357).

The diversity hierarchy can also be expressed as intraspecific diversity (genetic variation within species or populations, in other words, the genetic characteristics stored within the gene pool), interspecific diversity (community diversity or species richness and community structure), functional group diversity (useful for scaling up from population- to ecosystem-level analyses) and functional diversity among ecosystems within a landscape. Another way of expressing this is as alpha (α), beta (β) and gamma (γ) diversities. Alpha diversity is the species richness of samples representing communities, generally at the scale of 10^2 to 10^4 m^2. This is also known as within-habitat diversity. Beta diversity is between-habitat diversity or differentiation diversity, the amount of biotic change between units, for example between alpha diversity, the local scale, and gamma diversity, the landscape scale. Gamma diversity operates at the scale of landscapes, 10^6 to 10^8 m^2. Thus there is a nesting of each level within the higher level diversity (Shmida and Wilson, 1985). The approach to studying these levels, in terms of interpreting their determinants, differs according to the philosophical bases for allied disciplines such as classical biogeography, quantitative phytosociology or community analysis, and theoretical population ecology. Each has its own manner of interpreting the mechanisms causing diversity. Classical biogeographers stress historical factors such as migration, isolation and speciation in their interpretation of diversity. Their level of study tends to be gamma diversity. This can be seen in the historical ecology interpretation of the flora of the Mediterranean outlined in Section 5.1. Quantitative phytosociologists (see Box 6.1) assess diversity as dissimilarity or 'ecological distance' between community samples. They stress the importance of environmental determinants – beta diversity. Population ecologists focus on the role of species interactions, such as competition and predation, especially at the small scale and therefore at the level of alpha diversity.

Some researchers argue for another hierarchical level beyond that of the landscape and recognise delta (δ) diversity or the functional diversity among ecosystems within a landscape (Blondel and Aronson, 1995) and its importance for conservation. As a result there has been a move towards the preparation of Red Lists of Endangered Landscapes and the creation of Green Books for Landscape Conservation. A case study of this has been carried out in western Crete (Naveh and Lieberman, 1994).

7.2.2 Functional biodiversity

The role of functional diversity in ecosystem processes has led to an interesting debate regarding the possibility of ecological redundancy – the recognition that some species may not be as important in the operation of ecosystem processes as others. Functional diversity therefore examines the role of an organism or set of organisms within an ecosystem and the relative abundances of these functionally different kinds of organisms. Functional groups, or guilds, are sets of species which perform similar functions or roles – the way in which they regulate ecosystem processes, i.e. they are ecological equivalents (Walker, 1992). One such group are the nitrogen-fixing species of the Fabaceae family. This family contains more than 18 000 taxa in over 600 genera but not all are nitrogen fixers; the only two native Mediterranean leguminous trees, the carob (*Ceratonia siliqua*) and Judas (*Cercis siliquastrum*) trees are not nitrogen fixers. There is, however, a large proportion of herbaceous and shrubby legumes, both annuals and perennials, which are nitrogen fixers, such as *Astraglus* (milk vetch), *Cytisus* and *Genista* (both types of broom), *Medicago* (medick) and *Trifolium* (clover). These are generally present in early post-disturbance, or successional, stages. Blondel and Aronson (1995) examined the contribution of legumes to four regional floras within the Mediterranean – the Hérault in southern France, Catalonia in northeastern Spain, Tunisia and Israel. In the more arid regions of Tunisia and Israel legumes accounted for 12.5% and 14.9%, respectively, of their mediterranean-type floras compared with about 9% in both the Hérault and Catalonia. Herbaceous perennial legumes are especially abundant in the floras and so presumably figure prominently in nitrogen cycling. Grasses are also an important functional component of ecosystems but have more than one role. Some of their importance lies in their nutrient uptake. It is also possible that they form mutualistic associations with some legumes and together these play their most important role in early and especially middle stages of plant community succession, or development. This occurs after the initial colonisation of disturbed sites but before full tree cover is established (Blondel and Aronson, 1995).

Functional diversity assumes that not all species are created equal; some can be recognised as determinants, or 'drivers', of the ecosystem of which they are a part, while others are 'passengers', with the loss of passengers causing little difference to the functioning of the ecosystem. A key question is therefore which kinds of species are the drivers, or keystone species, and so most important to ecosystem functioning? If this can be answered then the implication is that some species or groups of species are less important in terms of ecosystem functioning as other species can substitute for them. It then follows that redundancy of species or sets of species might exist. This could mean that ecosystems have a built-in buffering capacity to compensate for species loss. However, it is likely that a threshold exists beyond which species loss will lead to a shift to different levels of system functioning and altered feedback systems. In other words it is the nature of the interactions between organisms which determines the stability of a community rather than simply the number of species present and their apparent function at any one time. Consequently we must not be complacent about the loss of one or a few species based on an assumption that other species can provide a substituting role.

The recognition of functional diversity, and therefore ecological redundancy, is also only based on ecological principles (Walker, 1992), not on other concerns for the conservation of biodiversity, such as the commodity values of certain species (for example for their as-yet-unknown pharmacological properties), amenity values whereby actual values are attached to certain species (generally large mammals) which attract tourist revenues and, not least, the moral reasons for conservation of all species. While recognising functional diversity removes the emphasis in conservation away from the approach that all species are equal and replaces it with one based on functionally different kinds of organisms, there are many inherent dangers in the philosophy. What happens if an apparent passenger species at one scale turns out to be an infrequent driver? A species' significance may only become apparent under a certain set of infrequent environmental conditions, difficult to observe without the long-term study of the component species and the functioning of their ecosystem. The recognition of functional diversity needs considerable research (Walker, 1992) but has been a part of studies of the function of biodiversity in regions with a mediterranean-type climate (Blondel and Aronson, 1995).

7.2.3 Keystone species as indicators of diversity

Related to the concept of functional species are those that can be referred to as keystone species. These are species with no ecological redundancy – indeed, they are pivotal species for ecosystem maintenance upon which the diversity of a large part of a community depends (Noss, 1990). Their demise would lead to a decline in the number of other species within that community, leading to a loss in biodiversity. Identification of keystone species may help effective ecosystem management.

For the Mediterranean Basin the identification of keystone species is beginning (Blondel and Aronson, 1995). No obvious keystone birds have yet been recognised, but some plant species have been suggested. These are nitrogen-fixing legumes, such as the herbaceous annuals and perennials referred to in the previous section. They are important in early to middle stages of post-disturbance vegetation dynamics but are absent from the later stages of woodland. It is argued that without their presence tree cover could not become established (but see Section 7.3 for a discussion on vegetation succession).

No Mediterranean mammals have been identified as keystone species in terms of their role in ecosystem functioning although some may take on this role because of their economic as well as ecological importance (Blondel and Aronson, 1995). The wild boar (Sus scrofa), for example, is a popular hunting target (Plate 7.1). Consequently some forest areas provide a higher economic return from hunting than from traditional forest products such as timber. The protection of boar together with other game species then aids the protection of other animal and plant life and so promotes species richness. This has been the case with much of the woodland in the Iberian Peninsula, known as dehesa or montado, which is examined further in Sections 8.5 and 9.3.4.

As a final identification of a group of Mediterranean keystone species, Blondel and Aronson (1995) mention earthworms. Species richness is high in Mediterranean earthworms, with 150 identified species in southern France compared with only 30 species in northern France. Their contribution to ecosystem activity lies in their physical effects (soil aeration, water drainage, bioturbation), their biogeochemical effects (litter decomposition, phosphorus and nitrogen recycling, balancing pH and maintaining the carbon/nitrogen equilibrium), their maintenance of forest diversity (through promoting germination and growing conditions of trees and herbaceous plants) and their role as a source of protein for many predators. Up to 200 species of birds and mammals feed on earthworms in southern France – this is one-third of the total species richness of these two groups at the scale of southern France. Yet earthworm diversity is declining and their communities are becoming

Plate 7.1   Wild boar, *Sus scrofa*, in S. Rossore nature reserve, Tuscany, Italy. © M. Biancarelli/
Woodfall Wild Images.

impoverished. This has been attributed to, among other factors, logging, litter
removal, heavy metal pollution, and acidification of soils through acid rain
and afforestation with conifers. As well as their role as ecosystem keystone
species, their disappearance has possibly contributed to an increase in soil
erosion and the occurrence of severe floods, especially in agricultural areas
where copper is widely used as a fungicide. Some species of earthworm, such
as those of the genus *Sclerotheca*, are sensitive to heavy metal pollution from
copper, zinc, lead, molybdenum, cadmium and iron.

7.2.4 Other indicator species for biodiversity

Keystone species are only one type that might warrant special conservation
effort. Four other categories of indicator species have been identified –
ecological indicators, umbrella species, flagship species and vulnerable species
(Noss, 1990). Ecological indicators are those species in which fluctuations in
population numbers might signal the effects of perturbations on a number of
other species which have similar habitat needs. It is possible that amphibians
may be good indicators of environmental change (Márquez and Alberch, 1995).

Umbrella species are those which require a large area for survival. They are
found at the higher trophic levels; in other words, they are large mammals and
birds of prey. Given a sufficiently large area of protected habitat their survival
may bring many other species into protection. For example, the brown bear
(*Ursos arctos*) in Abruzzo, Italy, needs a total area of 1500 km² for foraging
(Zunino, 1981). Studies in Cabañeros Natural Park in central Spain have also

indicated that conservation programmes established for the black vulture (*Aegypius monachus*) and Spanish imperial eagle (*Aquila adalberti*) will improve the conservation value of the park for vegetation and other bird species (Abbott *et al.*, 1992).

Flagship species are those which are popular and charismatic – generally photogenic. In the Mediterranean, possible candidates include the loggerhead turtle (*Caretta caretta*) and the green turtle (*Chelonia mydas*). Both breed regularly on the beaches of Cyprus, Greece and Turkey and their population numbers are threatened particularly by tourism development and incidental catches in fishing nets. As charismatic species they are the focus of several conservation groups (Poland *et al.*, 1996). Other possible flagship candidates are cetaceans. The governments of Italy, France and Monaco signed a treaty in 1999 to create a whale sanctuary between the French Côte d'Azur, Monaco, the Ligurian coast of Italy and the islands of Corsica and Sardinia. This commits the countries to intensify actions against sources of pollution at sea and from the coast. The area is a feeding ground for thirteen different cetaceans, including the pilot whale (*Globicephala melaena*), common dolphin (*Delphinus delphinus*), striped dolphin (*Stenella coeruleoalba*), bottlenose dolphin (*Tursiops truncata*) and Risso's dolphin (*Gramous griseus*). Any success in the conservation of loggerhead turtles, dolphins and whales will also help the conservation of other species found within their habitats – both marine and coastal.

Vulnerable species are those which face extinction through rarity and/or being genetically impoverished. This includes many endemic species (see Section 7.2.6). Rarity may result from persecution through hunting, loss of habitat or habitat fragmentation through land-use change. Population numbers can also depend on the availability of patchy or unpredictable resources; some species may have low rates of reproduction and may be extremely variable in population density. The category 'vulnerable species' is defined by the International Union for the Conservation of Nature and Natural Resources (see Table 9.1) and vulnerable species may be keystone species, flagship species or umbrella species. Indeed, some species may be regarded as both keystone and flagship, or flagship and umbrella species, etc. Issues surrounding the conservation of vulnerable and endangered species are examined further in Chapter 9.

7.2.5 Structural diversity

Yet another way of analysing the diversity of an ecosystem is to examine its structure (Noss, 1990). At the level of the regional landscape (see Figure 7.2) structure includes the size of an area of homogeneous habitat, its fragmentation and the size and number of its patches, and the configuration of any connections between patches (sometimes referred to as wildlife corridors). At the community-ecosystem level structural diversity may be represented by the community's architecture or the arrangement of the biomass into layered forms (leaf density and layering at different heights), canopy openness and the proportion and spacing of gaps in the canopy. It also includes the distribution of physical features within the community such as rocky outcrops, water courses and sources of food or other resources. Structure is therefore complex but is important in that it has a strong influence on the animal life that inhabits a site, community or region, as it provides the environment in which animals feed, move around, seek shelter and breed. As well as variations in spatial structure, it can change over time – see Section 7.3 on ecosystem development. Recognising the importance of structural diversity is implicit within plans for landscape conservation.

At the population-species level an analysis of structure includes variables such as microscale barriers to dispersal, the distribution of species within their community, population structure (age/sex ratios) and habitat variables. At the genetic level structure refers to a range of genetic characteristics such as heterozygosity (a measure of genetic variation), chromosomal polymorphism (the existence of one or more chromosomes in alternative structural forms within the same species) or phenotypic polymorphism (the existence within a species of more than one form of physical appearance). An example of this comes from a detailed study of the genetic variability within *Thymus vulgaris* (thyme). Considerable intraspecific variability in aromatic oil content occurs between individual plants growing short distances apart. This appears to be caused by slight changes in the growing environments of these individuals (Gouyon *et al.*, 1986).

7.2.6 Rarity and endemism

Diversity also includes the abundance or rarity of species and the way in which these are distributed and/or restricted in their range. Species restricted to a particular area are known as endemic, and there has been a long interest in endemicity. Areas with an exceptional concentration of species and high numbers of endemics, of either plants or animals or both, may be referred to as diversity hotspots (Myers *et al.*, 2000). Interest in these regions ranges from the search for explanations as to their existence to their importance in conservation. It may be that endemic species are those most endangered or threatened with extinction and therefore geographic rarity is sometimes used as a criterion for conservation. However this is not always the case. In the nutrient-poor heathland habitats of southern Spain plant species richness is low and there is a high proportion of endemic species. The endemics are not apparently endangered, however, because they are locally abundant (Ojeda *et al.*, 1995). Local abundance does not make the region any the less valuable for conservation but to assess its true value a combination of diversity indices need to be examined, instead of relying on just one.

Globally, 25 terrestrial diversity hotspots have been identified (Myers *et al.*, 2000), five of which are the regions with a mediterranean-type bioclimate. Sometimes the Mediterranean Basin has been excluded from hotspots analysis on the grounds that it covers too large an area, but detailed analysis of the region has identified hotspots or red alert areas within the Mediterranean, including the islands of the Canaries and Madeira – sometimes referred to as Macronesia (Médail and Quézel, 1997).

Table 7.1 Plant diversity in different regions with a mediterranean-type climate

Mediterranean-climate region	Area (km²)	Approx. no. of species	No. species per 1000 km²	Approx. no. endemic species	Percentage of endemic species
Mediterranean Basin	2 300 000	25 000	10.8	12 500	50
California	324 000	4 400	13.7	2 140	48
South Africa	90 000	8 500	95.5	5 860	68
Chile	140 000	2 400	17.1	1 450	60
Southwest Australia	310 000	8 000	25.8	6 000	75

Sources: Cowling *et al.* (1996); Médail and Quézel (1997).

In terms of species richness the number of vascular plants is high in the Mediterranean. The region is estimated to hold about 25 000 species of flowering plants and ferns in an area of about 2.3 million km². This can be compared with just 6000 species in non-Mediterranean Europe, an area of 9 million km² (Quézel, 1985). However, a more useful comparison might be with other areas of the world that have a mediterranean-type climate. These are compared in Table 7.1, though caution is needed in interpreting the data, especially with calculations based on area. As an area decreases in size, the number of species per thousand kilometres squared and the percentage of endemics nearly always increase. Another problem rests with the fact that the figures are not precise – floristic inventories for any region are difficult to make, in part dependent on different regional definitions and because floristic inventories are usually made with reference to administrative boundaries rather than geographic or vegetation boundaries. Different studies have produced different estimates of species richness and endemism (see for example Cowling *et al.*, 1996; Myers *et al.*, 2000). Nevertheless a crude estimate of species richness reinforces the floral diversity of regions with a mediterranean-type climate. Approximately 15% of the world's total number of plant species are found in regions with such a climate (Myers *et al.*, 2000). All five regions with a mediterranean-type climate are threatened to varying degrees by species loss through past and current human activity and potentially through anthropogenically-induced climatic change. Consequently a clear understanding of the role of biodiversity in the functioning of mediterranean ecosystems is needed before too many further losses occur.

Within the Mediterranean Basin two major centres of biodiversity have been identified – a western centre, including the Iberian Peninsula and Morocco, and an eastern centre including Greece and Turkey (Médail and Quézel, 1997). Nine sectors are distinguished within these two centres as having the greatest number of endemics which appear to be most endangered (see Figure 7.3). These are the High and Middle Atlas mountains, the Betic-Rif complex in the Iberian Peninsula and North Africa (grouped together because they were linked until the end of the Tertiary period), the Maritime and Ligurian Alps, the Tyrrhenian islands (the Balearics, Corsica, Sardinia and Sicily), south-

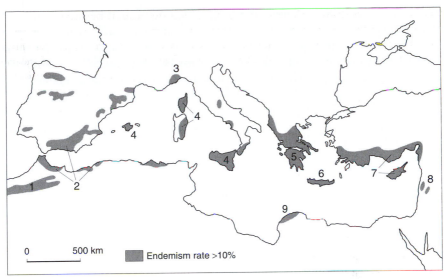

1 High and Middle Atlas Mountains
2 Betic-Rif complex
3 Maritime and Ligurian Alps
4 Tyrrhenian Islands
5 Southern and Central Greece
6 Crete
7 Anatolia and Cyprus
8 Syria-Lebanon-Israel
9 Mediterranean Cyrenaica

Figure 7.3 Diversity hotspots in the Mediterranean region. After Médail and Quézel (1997).

ern and central Greece, Crete, Anatolia and Cyprus, the Mediterranean parts of Syria, Israel and Lebanon, and the coastal region of Libya, Cyrenia. Within these hotspots rates of endemism may be as high as 50%, for example in the Betic Cordillera of Spain. Other mountainous areas with rates of endemism higher then 20% are the Rif in Morocco, the Middle and High Atlas ranges in North Africa, the Serra da Estrêla in Portugal, the mountains of southern Greece, Crete, the Taurus range in Anatolia, the Troodos mountains of Cyprus and the summits of the Lebanon and Anti-Lebanon. Médail and Quézel also include the islands of the Canaries and Madeira as Mediterranean hotspots.

7.2.7 Explanations for patterns of species richness

Various hypotheses have been proposed to explain plant and animal species richness in the five mediterranean-climate regions, including historical and palaeogeographical factors, the consequences of environmental changes, ecological stress, competition between species, diversity of habitats and, last but not least, disturbance. These factors operate at a range of scales and reinforce one another. This can lead to confusion surrounding the competing ideas but shows that explanations are linked both temporally and spatially. Here the explanations are explored starting at the longest and largest scales and moving towards shorter and smaller scales – in other words, from regional diversity patterns to local or alpha diversity.

Historical and palaeogeographical factors are generally cited as reasons for diversity at the regional scale. For plants in the Mediterranean region we can identify those elements whose origins are described as palaeo- or neo-Mediterranean, for example the flora which developed before or after the establishment of a mediterranean-type climate approximately 3 million years ago. We can also identify elements described as tropical and non-tropical, in terms of their origins in different floral realms (see Section 5.2). In addition, the consequences of environmental and particularly climatic changes are important. These partly determine whether exchange of plants and animals is possible between different regions as climates and the configuration of biogeographic regions change. For example, during the Quaternary glaciations of northern Europe, the Mediterranean region generally experienced aridity. As the Mediterranean became isolated from the north, floral exchange between it and northern Europe was not possible but it did provide refuge areas for temperate forests trees (see Section 5.2.4). Movement along plate boundaries and tectonic uplift have also helped to create isolated populations within which speciation has occurred, promoting endemism and species richness as, for example, the high species richness of the Italian and Iberian peninsulas.

Diversity changes along habitat and geographical gradients reflect geographical vicariants and the evolution of habitat specialists. In other words, evolution occurs with species adapting to particular environmental constraints such as summer drought. One consequence might be that in all regions with a particular environmental constraint, organisms will be selected for similar structural and functional attributes. This is known as convergent evolution (see Box 7.1). Evolution also occurs as a result of the presence of barriers to dispersal such as mountains separating coastal lowlands from the interior, the geographical arrangement of peninsular areas and the existence of seas separating islands from each other and from mainland areas. In the Mediterranean region

---

Box 7.1   Convergent evolution in mediterranean–type ecosystems

The theory of convergent evolution predicts that given similar environments, organisms will be selected for similar structural and functional attributes. A comparison of these attributes in the mediterranean-type ecosystems of South Africa, Australia, California, Chile and the Mediterranean Basin reveals many apparent similarities. This raises the issue of evolutionary convergence and non-convergence of their ecosystems (di Castri and Mooney, 1973; Cody and Mooney, 1978).

Floristically the similarities between the regions are low; they share few species in common and there is also dissimilarity at the generic and family level. However, a review of the five regions suggests that plant evolutionary strategies are in general comparable (Cody and Mooney, 1978). Although it is an over-simplification some of the selective forces, their consequences and the evolutionary strategies of mediterranean-climate shrubs are modelled in Figure 7.4. While summer drought is a major selective force, though to

(continued)

(continued)

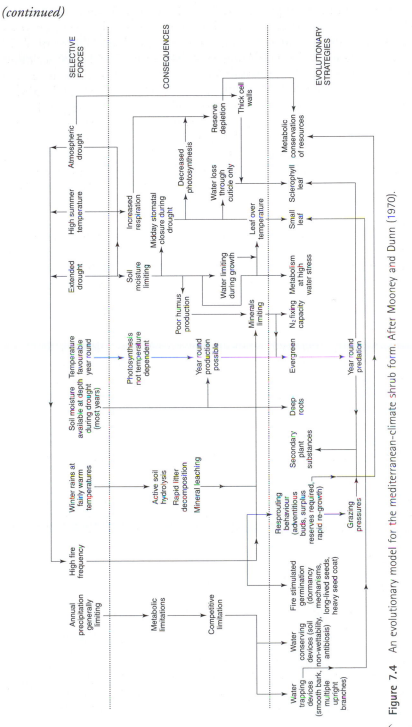

Figure 7.4 An evolutionary model for the mediterranean-climate shrub form. After Mooney and Dunn (1970).

(continued)

(continued)

varying degrees, other selective forces include grazing (especially around the Mediterranean Basin) and fire. Since the model was proposed several studies have attempted to assess its applicability.

Many plant species in all five regions show adaptations to fire and are able to exploit it for population expansion (see Section 5.5). Three particular mechanisms are used: light-stimulated germination, heat-shock and the chemicals generated in smoke, referred to as charred-wood-induced germination. Smoke is important for the germination of several fynbos species from South Africa and in the Californian chaparral. However, charred-wood germination has not been reported for any species in the Mediterranean basin. Keeley and Baer-Keeley (1999) therefore used laboratory germination tests to see whether species from phrygana communities in southern Greece responded to smoke, and to light and heat-shock. Only one species out of 22 was stimulated to germinate by charred-wood – *Lavandula stoechas* (lavender). Heat-shock was important, especially in species of *Cistus*, while response to light ranged from neutral in *Cistus* species, through positive in *Misopates orontium* (a member of the foxglove family, Scrophulariaceae) to light-inhibited in the *Allium* species. Their conclusions were that Mediterranean Basin flora is relatively depauperate in charred-wood/smoke-stimulated germination and that this represents a lack of convergence with other mediterranean-type ecosystems. Although there is a similarity between the five regions in terms of response to heat-shock, this itself might not be a feature of convergent evolution but rather of parallel evolution. Species of Fabaceae (the pea family) and Cistaceae (rockrose family) are common in both California and the Mediterranean. Keeley and Baer-Keeley further speculate that in the Mediterranean the long history of intensive land use may have selected for more opportunistic type colonisers, in other words those in which germination responds to light and heat which occur with disturbance.

An examination of the ancient lineages of mediterranean flora (those that evolved before the establishment of a mediterranean-type climate) also suggests that taxa which have persisted into Pleistocene and modern times have retained their tropical characteristics, not evolving special adaptations to a mediterranean climate (Herrera, 1992).

Convergence would not, of course, be confined to plants. Birds, for example, might share attributes of shape, wing length, bill characteristics etc. in relation to foraging techniques between the five mediterranean-type ecosystems. However, a study comparing bird communities in France, Chile and California, which were then compared with non-Mediterranean France, found no evidence of convergence (Blondel *et al.*, 1984). No more similarities were found among the mediterranean-ecosystem bird communities than between any of them and the control non-Mediterranean group. Nevertheless an interest in the apparent convergence of morphological and structural attributes between the five mediterranean regions remains.

there is marked heterogeneity of landscape over very short distances – differences in relief, slope, climate, rock type and soil type – and these have resulted in biogeographical, taxonomic and genetic diversity. This can be seen on Mount Carmel in Israel where there are dramatically different populations of European and African faunal and floral elements on opposing north and south facing slopes (Naveh, 1995). Though separated by only a few hundred metres, on the exposed and open, and therefore drier, southern slopes there is distinctly higher genetic and species diversity. These differences occur in such unrelated taxa as plants, insects, snails, reptiles, birds and small mammals.

Variations in diversity at the genetic level within a single species can also be explained by climatic variability. A study of 107 genotypes from 27 wild emmer wheat populations in the Mediterranean part of Israel shows that significant genetic variation occurs between the populations, explicable in terms of water availability, radiation receipts and temperature. The highest photosynthetic efficiency and the greatest genotype variability are found in the drier areas with a more marginal mediterranean-type climate. In these areas the length of the growing season is shorter and there is greater unpredictability in seasonal rainfall. Moisture stress can be severe. These conditions have favoured the development of more drought-tolerant genotypes (Naveh, 1995).

At the local scale the frequency of disturbance appears to be important. In some mediterranean-climate regions fire is the main form of disturbance, especially in the Cape region of South Africa and in southwestern Australia (Naveh, 1994). By comparison, fires are less frequent around the Mediterranean, but locally they are common and are increasing in frequency. Their effect is reinforced by grazing and browsing by cattle, sheep and goats, which are traditional land uses in the Mediterranean Basin. Fire and grazing allow the coexistence of short-lived and long-lived plant species. The result is the maintenance of diversity and the recognition of the need to maintain disturbances, rather than suppressing them. This gives rise to what are recognised as perturbation-dependent landscapes. Both of these factors are examined in more detail in Sections 7.3.2. and 7.4.

7.3 Ecosystem development – to succeed or not to succeed? ____

Ecosystems are not static, unchanging entities but dynamic. As such, the diversity of a site changes over time. This may be over periods of thousands to millions of years, but exploring these is hindered by incomplete fossil records for different time slices. More observable are those studies of diversity changes over periods of decades or centuries. Repeated observations on some sites over periods of several decades have become classic studies of vegetation change but in non-mediterranean areas, such as at Glacier Bay in Alaska. In most climate regions continuous or repeated scientifically based observations have only been made more recently. Some clues, however, are available from early photographs and paintings of landscapes, which can be compared with modern views. Rackham and Moody (1996) cite the landscapes drawn by the humorist Edward Lear on his travels to Crete in 1864 showing a less vegetated scene

Crete; between 1945 and 1989 coniferous forest grew in area by some 20% (Papanastasis and Kazaklis, 1998). This increase, however, was not necessarily a result of planning, more a consequence of changing socio-economic conditions. Since the 1960s cereal and vineyard cultivation on terraced slopes declined, to be replaced by more intensive cultivation of olive and citrus groves on the plains and newly created terraces; coniferous forest expanded onto the former terraces. At the same time grazing pressures were reduced – fewer goats and sheep were pastured on the plains and hilly areas – but the impact of grazing on the landscape was also being reassessed. There was a growing recognition that, in many Mediterranean regions, livestock numbers had long been in equilibrium with the grazing capacity of different vegetation communities, partly through the use of fire to control the density and spread of unpalatable woody plants and promote the growth of palatable vegetation.

Re-evaluations of Mediterranean landscapes lead us to question the extent to which these landscapes are a reflection of cultural impact. In other words, are bare limestones the result of anthropogenically induced soil erosion or instead the result of earlier erosion during the Pleistocene? Further questions follow on: if human activity is responsible, then which particular aspects of human activity are most likely to have resulted in landscape change, and are the same types of activity occurring now as in the past? In other words, what distinctions can we make between 'traditional' land-use practices and more modern, intensive agriculture? We also need to determine the extent to which landscapes can be described as resilient or fragile, stable or destabilised. The answer to most of these questions lies in land-use changes dating back thousands of years and therefore we also need to assess the evidence (archaeological and palaeoecological) for anthropogenically derived landscape change.

8.1 Pre–Neolithic landscapes

The beginning of the Holocene saw major changes in the vegetation of the Mediterranean region, compared with that of the late glacial period. Prior to about 14 000 years BP, much of the northern Mediterranean was dominated by open steppe communities of *Artemisia* (wormwood) and Chenopodiaceae (the goosefoot family), indicative of aridity. Pollen diagrams from early to mid-Holocene sediments show that the herb-steppe communities were then replaced by subhumid forests, mostly broadleaved deciduous trees or conifers, or by savanna in drier places (see Section 5.2.4). Such vegetation changes were a response to a global warming and to wetter conditions in southern Europe. However, the evidence for climatic changes and their timing is varied across the region. Some indications come from lake-level changes, as outlined in Section 2.7. In the early Holocene, lake levels suggest for several places that conditions were wetter than either previously or the present. Wetter conditions are suggested for the western Mediterranean from about 12 000 to 5000 BP, when aridity abruptly set in. In contrast in the eastern Mediterranean, lake levels were not at their maximum until around 6000 BP with the transition to contemporary drier conditions taking place gradually from about 5000 BP.

Pollen records indicate the spread of drought-adapted taxa in response to this increasing aridity. Typical 'mediterranean' evergreen trees and shrubs spread gradually westwards – from the Levant, southern Turkey, Crete and southern Greece – about 10 000 years ago (see Figure 8.1).

In North Africa pollen evidence from a variety of sites suggests that aridity prevailed from 18 000 years ago until about 6000 years BP (Rognon, 1987) with strongly seasonal variations in rainfall intensity causing intense erosion, especially earlier in this period. The transition to wetter conditions in mid-Holocene times was not synchronous across the Maghreb. For example, the sedimentary record at Tigalmamine in the Moroccan Middle Atlas mountains shows a marked increase in both evergreen and deciduous oak forests from 8500 years BP (Lamb *et al.*, 1989). This earlier date might reflect a more immediate response to climate change than elsewhere. The change in pollen record is coincident with a lake-level rise implying a rapid response of vegetation to climatic change, attributable to a close source of seeds for the re-establishment of forest – in other words, no migration lag existed between climate and vegetation response. Indications in the Tigalmamine pollen record of human activity begin some time around 5000 years BP. Consequently it is difficult to distinguish climatic signals after that date – whether or not there was greater aridity in the late Holocene compared with earlier. A review of sites across the Maghreb suggests that there was an increase in aridity in Tunisia from about 5000 BP but that this was not matched in northern Algeria and the coastal and mountainous regions of Morocco (Allen, 1996).

In order to assess whether the early to mid-Holocene existence of drought-adapted vegetation types around the Mediterranean Basin might owe as much to human impact as to climate change, Roberts (1998) examined pollen records from two other regions with a mediterranean-type climate, California and southern Australia. He wanted to determine whether sclerophyllous evergreen communities were present before European contact or only became dominant with the relatively recent introduction of agriculture. Both chaparral scrub (in California) and mallee heath (in Australia) showed a pre-colonial existence. Human activity therefore might not be a necessary precursor for sclerophyllous scrubland. If this argument holds true for the Mediterranean Basin, then the early Holocene settlers might have found a broad landscape of subhumid forest (later replaced by maquis-type vegetation communities) or alternatively a savanna-type landscape of trees with maquis. The pollen evidence is difficult to interpret in terms of closed or open woodland (see Box 8.1); inevitably, local variations existed. Valley bottoms, coastal plains and steep hillsides each had their own communities and were exploited by hunting–fishing–gathering communities and later by Neolithic farming communities.

Subsistence economies rely on manipulation of natural ecosystems without transforming them (Roberts, 1998). An example of such resource exploitation by hunter–fisher–gatherers comes from Portuguese Estremadura, north of the Tagus river (Aura *et al.*, 1998). Archaeological finds show that the key game species for the period 13 000 to 8000 years BP were red deer (*Cervus elaphus*) and rabbit (*Oryctolagus cuniculus*). From about 10 000 years BP the importance

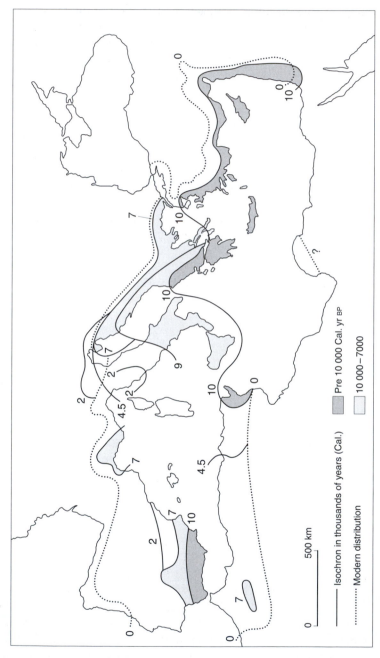

Figure 8.1 Holocene spread of summer-dry woodland in the Mediterranean Basin. After Roberts (1998).

Box 8.1 Pollen indicators of anthropogenic activity

When looking for signs of human impact on vegetation in the early Neolithic, a common starting point is often the changing ratio of tree pollen (or arboreal pollen, AP) to non-tree pollen (non-arboreal pollen, NAP). The assumption is that woodland clearance leads to replacement (or at least reduction) of tree cover by open ground or herbaceous vegetation, and that this is reflected in pollen diagrams derived from sediments retrieved from the region. Unfortunately the interpretation of pollen diagrams in this way is rarely simple. Pollen analysis, or palynology, is founded on the assumptions of uniformitarianism, in other words that past events can be interpreted in terms of present-day processes. Thus we need to make assumptions that plants and vegetation communities 'behaved' in the past in the same way that they do today, that pollen production in the past was similar to that of present-day conditions, and that we can find modern analogues (for example, the composition of vegetation communities) which can be used to describe past communities. We therefore need to have a good understanding of factors such as how much pollen different taxa produce, how different plants are pollinated, and how different vegetation communities and individual components respond to such human activities as clearance of woodland, cultivation, grazing and use of fire. These issues can be illustrated by examining some of the taxa which have been used as indicators of anthropogenic activity in the eastern Mediterranean and Near East.

It might be assumed that the beginning of agriculture in the Near East, in terms of the domestication of wheat and barley, would be clearly indicated in pollen diagrams with the appearance of cereal-type pollen. This is not so. Barley and wheat are both members of the grass family (Gramineae) and some wild grass species are difficult to distinguish from cereals (Bottema and Woldring, 1990). In addition, cultivated wheat and barley are self-pollinating and a large part of their pollen remains on the spikelets. This means that the pollen is poorly dispersed, and so these cereals tend to be under-represented in the pollen rain (the composition and amount of pollen that is collected in pollen traps, or becomes deposited in sediments). There is also a high incidence of wild grasses with cereal-type pollen indigenous to the Near East and eastern Mediterranean. Thus their occurrence in pollen diagrams need not indicate farming. It is not until such pollen is found at sites outside its natural range (as in northern Europe) that its presence is unambiguous and can be regarded as a primary anthropogenic indicator.

Primary indicators are those species which are cultivated in fields, gardens and orchards (Behre, 1990) but, as with the grasses, the evidence given by these species is often limited. The occurrence of olive (*Olea*) in pollen diagrams dating to the Neolithic is considered to be derived from wild olives, because archaeological evidence suggests that it was not cultivated until the Bronze Age. Even after that time, the presence of olive may not indicate its local cultivation. As it is wind-pollinated, it is well represented in modern

(continued)

(continued)

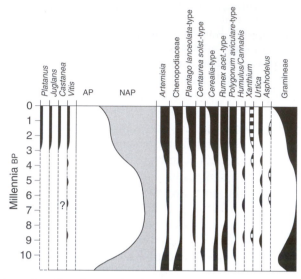

Figure 8.2   Schematic pollen diagram showing a selection of pollen curves indicating human influence in Greece. After Behre (1990).

pollen rain outside the area in which it is grown (Bottema and Woldring, 1990). The geographical range of the vine (*Vitis*) stretches from Greece to northern Iran, but it was, and is, widely grown outside this region. A natural climber which is usually pruned and trained into a shrub, its pollen production is very low throughout the varying vegetation zones of the eastern Mediterranean. It is therefore accepted that when its occurrence in pollen diagrams is greater than only 0.1%, it should be regarded as growing in the vicinity of the pollen site (Bottema and Woldring, 1990).

As the value of primary anthropogenic indicators in pollen diagrams is often limited, secondary indicators may be of help in tracing different human activities in the past. These are species not intentionally cultivated but favoured in various ways by human activity, for example 'weeds' or ruderal species which take advantage of the disturbed ground often created by cultivation. A classic indicator in northern Europe is ribwort plantain (*Plantago lanceolata*), which thrives on disturbed ground. As a pioneer species it is pre-adapted to survive in habitats disturbed by the spread of Neolithic agriculture. The same story is apparent in the Near East. Here it is sometimes taken as the most reliable indicator species for grazing (Behre, 1990). In pollen diagrams from Greece and Turkey, percentages of arboreal pollen decline from about 4000 BP (see Figure 8.2), while percentages of *Plantago lanceolata*-type pollen increase, together with other 'weeds' characteristic of open vegetation, such as wormwood (*Artemisia*), goosefoot (Chenopodiaceae), knapweed (*Centaurea*)

(continued)

(continued)

and docks (*Rumex*). Cereal-type and *Olea* pollen also increase at this time and the following trees appear from about 3000 BP: walnut (*Juglans*), sweet chestnut (*Castanea sativa*) and plane (*Platanus* spp.) – all likely indicators of deliberate planting or introduced into the area.

The value of 'weeds' is, however, explored by Blumler (1996), who cautions that they should not always be regarded as disturbance indicators. He argues that they play this role because in traditional notions of vegetation succession, weeds are early successional taxa. Theories of vegetation succession deem disturbance events to be rare, occurring early in a successional sequence or as a result of human activity. Blumler argues that in reality ecological studies show disturbance to be much more common and that most plant species are adapted to, or even require, some form of disturbance. Therefore, 'weeds' should not necessarily be regarded as disturbance indicators when other 'wild' (non-weedy) species are not, as ecologically they may in fact be. This view is not widely adopted in the interpretation of pollen diagrams, especially as it is a combination of indicators, rather than individual taxa, that should be used to deduce human activity, with corroboration from other disciplines.

Even when the ratio of tree to non-tree pollen is used to identify woodland clearance episodes its validity may not cross from one part of the Mediterranean to another. It is commonly accepted that the Balkans were well-wooded early in the Holocene (Willis and Bennett, 1995). Elsewhere this may not have been the case. Rackham (1998) cautions that it is difficult to distinguish forest from maquis communities when the same species are present in both as either trees or shrubs. In a maquis community, taxa commonly present as undershrubs or herbs are poorly represented in pollen diagrams. For example, undershrubs such as *Cistus* and species of the mint family (Lamiaceae) are poor producers; so too are some herbs that are insect-pollinated, such as asphodels (*Asphodelus*). This makes it difficult to interpret vegetation cover definitively as either forest or maquis, an important consideration when maquis is often regarded as a degraded, anthropogenic landscape. Modern ecologists often regard the presence of asphodels as indicative of excessive grazing. This could not be deduced from pollen diagrams.

The behaviour of trees in response to grazing and browsing is also important. The kermes oak (*Quercus coccifera*) may grow as a tree or a shrub. Both forms produce acorns and pollen, and a shrub may grow into a tree if it escapes browsing, or a tree become a shrub if browsing pressure is too great. This tree has been described as 'perhaps the most adaptable plant in the European flora. It can be bitten down to a few centimetres high and live indefinitely; it can also be a giant oak with a trunk 3 m in diameter' (Rackham and Moody, 1996, p. 64). The existence of *Quercus* in a pollen diagram cannot tell us about its growth form, and further, it is not possible to identify *Quercus* grains at the species level, though it is possible to distinguish between evergreen and deciduous forms.

of aquatic resources increased. Indeed, shell middens comprising estuarine and marine species are found at sites which would have been some 40 km inland from the coast. Aquatic resources were especially important to communities at Mesolithic sites along the estuaries of the Tagus, Sado and Mira rivers (about 8000 to 7500 years BP). Evidence suggests that up to 50% of the diet of these people was provided by aquatic food. Reliance on such a foraging subsistence along the lower reaches of the Sado and Mira rivers is believed to have continued until about 6000 years BP. Thus the last hunter–fisher–gatherers lived contemporaneously with Neolithic groups whose economy was based on agro-pastoralism. Neolithic groups had almost certainly colonised the calcareous hills of the interior by 6400 years BP (if not earlier, around 6800 to 6700 years BP), bringing agricultural skills from outside the region. Within the Iberian Peninsula the amplitude of ecological change after the Last Glacial Maximum was not as great as elsewhere in Europe, namely southern France, where climatic changes were quite marked. Consequently Mesolithic people in the Iberian Peninsula had a long period during which to diversify their subsistence strategies. Aura *et al.* (1998) note the significant degree of parallelism, with respect to this resource manipulation, between the different communities of the Peninsula, even though there is little knowledge about the interconnections of the people living in peripheral, coastal regions separated by inhospitable mountains.

8.2 Early agriculture

Mesolithic hunter–fisher–gatherers lived in many Mediterranean localities and an important question is the extent to which there was continuity or change between these communities and later Neolithic people. Did Mesolithic people gain knowledge of agricultural techniques as these spread from other knowledgeable groups, or was there independent invention at many different sites? Equally important is the question of whether Mesolithic sites were continuously inhabited by one group of people who developed/adopted new ideas and techniques, or whether sites were recolonised by newly arrived groups with these skills. These questions have driven much of the research into early agriculture in Europe and the Near East; agriculture has been, and remains, the most significant agent of environmental change as people have altered their immediate environment, with removal of natural vegetation leading in places to soil erosion.

A range of evidence exists to unravel the mysteries of early agriculture, including palynology, anthropology, archaeology, genetics and linguistics. All are used in an attempt to answer the questions of how, when, where and why people first domesticated plants and animals. There are some ambiguities in the use of terms such as domestication, agriculture, cultivation and husbandry, but each involves a degree of interaction between people, and plants and animals, with decreasing dependence on wild plants and animals and increasing dependence on domesticated ones over time (Harris, 1996).

A common assumption is that agriculture is more time- and energy-consuming than hunting–fishing–gathering. If this is so then an obvious question is why agriculture developed in the first place. A definitive answer is not available, but several hypotheses have been proposed, which Mannion (1995) divides into two groups: environmentalist and materialist. Environmentalism advocates environmental change as the stimulus for hunter–fisher–gatherer communities to change their means of food production – climatic changes between 14 000 and 10 000 years ago leading to a redistribution of plants and animals, including people, who, in southwest Asia at least, began to build settlements. Open-ground, light-demanding wild grasses would have spread as competition with trees and shrubs declined. Materialism advocates a need for new methods of food procurement because of population increase and a desire to abandon a nomadic existence for a settled way of life. Mannion also suggests an element of greed through a desire to produce surplus food and drink to ensure power and security. It is also likely that population increase led to the spread of people (demic diffusion) and ideas. Population movements could also have occurred as a response to climatic changes which made resources scarce and so reduced carrying capacity. Environmentalism and materialism need not therefore be mutually exclusive. A further explanation is the so-called 'dump-heap' hypothesis that mutual dependence had developed between certain types of plants and animals, and people. This is based on the idea that hunter–fisher–gatherers would have collected wild grain, some of which would have been discarded in middens. Here the seeds would have germinated in the organic rich environment, and then been harvested. Seeds with advantageous traits (see Section 8.2.1a) might have been selected preferentially and so, ultimately, domesticated. It is, however, unlikely that the reasons for plant and animal domestication will ever be known for certain.

8.2.1 Centres of origin

Agriculture developed independently at different times and places around the world. Centres of domestication for plants are commonly listed as the Near East, the Far East, South East Asia and the Pacific, and the Americas. For animals the important centres are Central America, South America, the Near East, the Indian subcontinent, South East Asia and the Ukraine (for the horse). For the Mediterranean region, therefore, the major centre of origin is the Near East from where agriculture spread to the north and west.

8.2.1a Plants

In the Near East, agricultural economies emerged between 10 000 and 8000 years ago. Until recently, the earliest dates for the domestication of plants (particularly wheat) were believed to have been from here, though it is possible that in China rice was domesticated earlier. The range of crops and early dates shows the importance of the Near East (Table 8.1). This region has become known as the Fertile Crescent and evidence for the emergence of agriculture

Table 8.1   The timing of common crop domestications

Crop	Common name	Approx. date of domestication, uncalibrated radiocarbon age (yr BP)
Hordeum vulgare	Barley	9800
Tritium dicoccum	Emmer wheat	9500
T. monococcum	Einkorn wheat	9500
Lens esculenta	Lentil	9500
Avena sativa	Oats	9000
Secale cereale	Rye	9000
Vicia faba	Broadbean	8500
T. aestivum	Bread wheat	7800
Olea europaea	Olive	7000

Source: Mannion (1999).

between 10 000 and 8000 years ago has come from over 50 archaeological sites. At that time reconstructions of vegetation (Figure 8.3) show forest to the north in Anatolia, bordering the Black Sea and in the Zagros mountains, subtropical woodland bordering the forested region and along the coast of the Mediterranean, and steppe grassland giving way to desert grassland towards the interior (Smith, 1995). With the transition to a warmer and wetter inter-glacial climate around 13 000 BP, growing conditions improved for wild grasses, such as barley and emmer and einkorn wheat. This area thus had a variety of vegetation communities and a range of climates which bordered and, in places, were akin to mediterranean-type climates. Today parts of the region are included as Mediterranean in bioclimatic maps (see Box 2.1).

Emmer and einkorn are two of several species of wheat brought into cultiva-tion. Through painstaking analyses of plant remains, including seeds, from archaeological sites throughout the Fertile Crescent, the wild ancestors of these two species are shown to have different geographical ranges (see Figure 8.3(a)) (Smith, 1995). Analysis of the DNA of wild and cultivated einkorn indicates that the most genetically distinct lines come from the Karacadag mountains of southeast Turkey (Heun *et al.*, 1997). The close genetic correspondence between the two suggests that wild einkorn is the ancestor of cultivated einkorn, both of which are found at archaeological sites in the region, and that initially they would have been cultivated alongside each other. Analysis of the mor-phology of plant remains at archaeological sites throughout the region shows changes from wild to domesticated plants, such as the shape and size of the grains, which became larger and plumper. The miniature stem which joins the grain to the plant, known as the rachis, also became tougher and less brittle (see Figure 8.4). Dating these morphological changes gives the timing of domestication and a clue as to how people were using the available plant resources both before and after domestication. For example, it appears that wild emmer and einkorn wheat were widely harvested before being domest-icated. For emmer wheat, the larger and plumper grains with a tough rachis

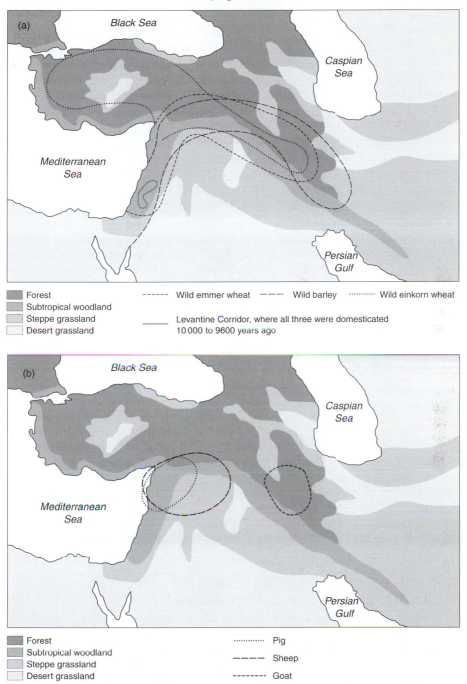

Figure 8.3 The Fertile Crescent: (a) the geographical ranges of the wild ancestors of emmer wheat, einkorn wheat and barley; (b) the areas of southwest Asia in which sheep, goats and pigs were domesticated. Adapted from Smith (1995) and Mannion (1999).

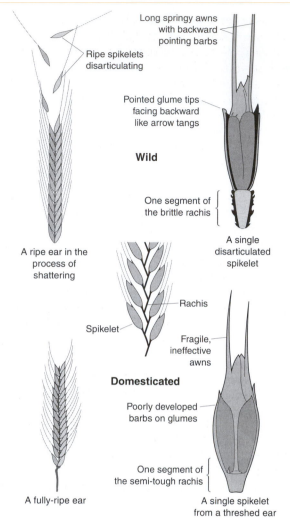

Figure 8.4 Wild and domesticated forms of einkorn wheat. After Smith (1995).

first appeared at the sites of Jericho, in the southern Jordan Valley, and Aswad near Damascus between 9800 and 9500 BP (Smith, 1995).

Domestication brought about an important difference in the response of the wild and cultivated plants to threshing. In wild wheat the ears are brittle and shatter on reaching maturity, so that the seed is disseminated close to the point where the plant grows. In contrast, cultivated wheat has non-brittle ears that stay intact after maturation (Figure 8.4). Early farmers must have selectively chosen individuals of those wild ancestors which showed this trait and harvested them, later sowing the seed for the following year's crop. In this way, domestication of those grains with harvesting and threshing adaptations which were more useful to people occurred. Smith (1995) suggests that this may have happened over a period of only three hundred years, from about 10 000 to 9700 BP. Even with a probability of one wild plant with this advantage out of every 2–4 million, such plants could have come to dominate in a short period of time. Cultivators would have had to harvest, by sickle or uprooting, those individuals with spikelets that stayed attached to the plant on maturity. The remaining plants with a brittle rachis would shatter and set seed. Some of this seed would be eaten by birds and rodents; some would germinate the following year. But quickly, seed sown from harvested individuals with non-brittle rachises would come to dominate, even though their survival depended on people for their reaping, threshing and sowing. Grain could now be stored from season to season and transported, as indicated by the fact that the earliest domesticated einkorn was found outside its natural range at Aswad near Damascus in the Levantine Corridor (Figure 8.3(a)) between 9800 and 9600 BP (Smith, 1995).

Wild emmer, einkorn and barley thrive in ecologically marginal areas such as open steppe, rather than on well-watered alluvial soils. However, they have more than adequately survived their relocation to sites more conducive to human settlement, and so become a reliable resource. Wild barley would have been harvested by early farmers and hunter–gatherers in the Fertile Crescent (see Figure 8.3(a)). Its domestication occurred alongside that of wheat in the same localities. These three cereals together with lentils, peas and flax played a major role in the early spread of Neolithic agriculture beyond the Near East (see Section 8.2.2). They occurred regularly in the Neolithic farming settlements that appeared first in Cyprus and Greece and subsequently in the Balkan countries and the middle and western Mediterranean Basin (Zohary and Hopf, 1993). The suggestion that domestication occurred over a short period in the area of the Fertile Crescent does not imply that transition from a hunter–gatherer economy to a sedentary one was as rapid elsewhere. Indeed, the speed with which settled agriculture replaced hunting, fishing and gathering will have varied from place to place, and is considered further in Section 8.2.2.

8.2.1b Animals

The Near East was not only a centre for plant domestication but for animals as well, although it is accepted that for some domesticates there were, worldwide, at least two centres of domestication. The wild ancestors of cattle, sheep, goats and pigs were aurochs, mouflon, bezoar goats and wild pigs, respectively; all were hunted in the Near East prior to domestication. Distinctions can be made between bones of wild and domesticated populations found at archaeological sites on the basis of parameters such as the proportions of bones of males and female animals, and the age at death. For example, it is argued that in goat herds managed for meat production there would be a large percentage of animals slaughtered late in their immaturity (when they had attained much of their adult size) and an adult breeding population in which females outnumbered males. Wild, hunted populations would have had a balance of adult males and females and fewer young goats (Smith, 1995).

8.9 Forest plantations

Trees have been widely used and planted around the Mediterranean Basin since prehistoric times (Le Maitre, 1998). The Carthaginians, Greeks and Romans used resins from pines for a variety of reasons. For example, pine resin has been added as a preservative to Greek wines, producing retsina, since the early Greek period and evidence exists for the cultivation of pines from Graeco-Roman times.

From the mid-nineteenth century reafforestation programmes have occurred in most Mediterranean countries, often as protection from soil erosion and for slope stabilisation. In particular pine and eucalyptus have been widely grown and undoubtedly some pine species have extended their natural ranges around the Mediterranean as a result of planting and invasion from plantations; *Pinus pinaster* has been particularly successful. This now covers some 2.43 million ha in Portugal, Spain and France. *P. pinaster* and *P. pinea* were planted along the coast of Tuscany in the nineteenth century and have since spread from their plantations (Barbero *et al.*, 1998).

Major afforestation programmes have been a feature of the twentieth century. Modern forest protection and management allows for multiple use of trees – not only is there soil protection but production of resin and timber. In addition there has been planting for aesthetics in attempts to recreate former extensive areas of forest and to spread forests into previously unforested regions. In Israel about 100 million trees were planted between 1948 and 1968 on some 40 000 ha, mostly of *Pinus halepensis*. Intensive site preparation and retention of runoff has allowed successful planting in the Negev desert where annual rainfall is less than 200 mm (Le Maitre, 1998).

Many afforestation projects have used pine species native to the region such as *Pinus nigra*, *P. pinea*, *P. sylvestris* and *P. uncinata* but only recently has detailed consideration been given to the ecology and genetics when selecting species for use – there is a need to alleviate possible negative effects which occur with certain species, such as their impact on soils and their possible inflammability. For example, *Pinus halepensis* produces serotinous cones which release seeds to germinate after a fire. It is one of the most inflammable and fire-prone pines, well adapted to frequent and intense fires. It may have extended its range into areas that were once forests of less fire-tolerant conifers. However, it is only successful if fire frequency is greater than once a decade – in areas where fires occur more frequently, the seed sources for *P. halepensis* may become exhausted with the creation of resultant shrubland (Agee, 1998).

In addition to pines, eucalyptus is widely planted. There are over 600 eucalyptus species, mostly native to Australia, but they have transformed many tropical and subtropical countries. As their leaves are held vertically rather than horizontally (an adaptation to intense radiation) light is reflected from them in a different way from most other trees (Mabberley and Placito, 1993). This gives a shimmering look to a eucalyptus-dominated landscape. Various species of eucalyptus were introduced to the Mediterranean from the early nineteenth century but were often not extensively planted until the 1960s, for example in Portugal. The most commonly encountered species is *Eucalyptus globulus*, the blue gum of Tasmania and Victoria. It grows fast, averaging 22 m height within ten years of planting and 50 m^3 of timber per hectare. By contrast *Pinus pinaster* takes 50 years to achieve the same height under ideal growing conditions and yields only 10 m^3 per hectare of timber. The rapid growth of eucalyptus and the multiple uses of its by-products explain its popularity. By 1990 afforested areas of eucalyptus covered some 500 000 ha of Portugal, accounting for more than 15% of total forest land (Mabberley and Placito, 1993). It is grown especially for paper pulp – its pale colour means that little bleaching is needed to produce high-valued 'white' papers. In addition, it is valued for firewood, oil and as forage for bees. As the European Union has a wood deficit, planting of eucalyptus is likely to continue around the Mediterranean region, despite the ecological problems associated with its cultivation. These centre around the potential risks of increased fire, effects on water availability, effects on soil nutrients, soil erosion and monoculture.

Eucalyptus is highly inflammable, in part because of its oil content and its dried leaves, which collect on the forest floor in summer. Although it is feared that eucalyptus plantations increase the risk of fire, there is, as yet, little corroborative data, especially as a high fire risk already exists in many Mediterranean areas. Nevertheless its growth as part of a monocultural system might exacerbate the fire risk. Monoculture potentially poses serious problems with respect to pest infestations. In Portugal there is evidence for serious infestations of fungal diseases and a beetle, *Phoracantha semipunctata*, which affects up to 12% of plantations (Mabberley and Placito, 1993). As eucalypts grow so fast there is concern that they use scarce water and lower the water table. This is less of a problem when they are grown in relatively wet upland areas, but the fear is recognised in lowland areas of Portugal, where they may not be grown within 20 m of cultivated land. Their rapid growth might also lead to nutrient depletion in the soil. However, the relationships between eucalyptus and soils are not fully understood, though there would appear to be a significant effect on pre-existing flora. Species richness under eucalyptus canopies is much reduced and plantations often replace vineyards, olive groves and pasture.

8.10 Landscape stability in the face of human activity _____

Chapter 4 included a discussion of the causes of soil erosion, whether anthropogenic, climatic or tectonic. Even if it is accepted that for much of the Mediterranean it is unwise to invoke largely climatic changes, as these are usually made with reference to those which occurred in northwest Europe rather than the Mediterranean, we do need to consider the question of landscape stability in the face of thousands of years of human activity. To what extent is stability the norm, or is it the case that land degradation, especially

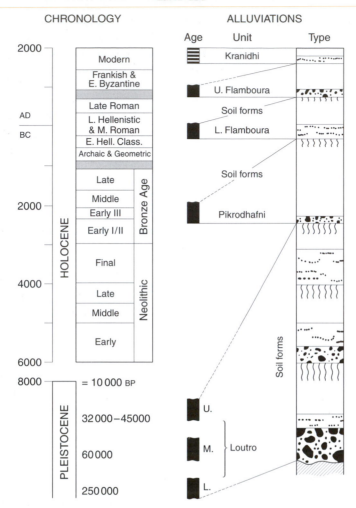

Figure 8.6 Chronology and stratigraphy of Late Quaternary alluvia and soils in the southern Argolid, Greece. Shading indicates 'dark ages'. Date and duration of each alluviation event is shown in the column labelled 'Age'. Broken bars indicate intermittent deposition; wavy terminations indicate uncertain age boundaries. Black cobbles indicate debris flows; strings of pebbles indicate streamflood deposits; blank areas represent loam. The length of wavy vertical lines is proportional to soil maturity and unit heights are roughly proportional to thickness. After van Andel *et al.* (1990a).

soil erosion, has been a persistent problem? One way of attempting to analyse this is through an evaluation and dating of episodes of soil erosion and stream aggradation. Such studies have been carried out in the Iberian Peninsula (Gilman and Thornes, 1987), southern Italy (Delano Smith, 1979) and Greece (van Andel *et al.*, 1990a). The results are relatively consistent, suggesting that periods of stability have been longer than periods of destabilisation. This is illustrated in Figure 8.6 for the southern Argolid (van Andel *et al.*, 1990a). Several phases of alluviation are shown for the Pleistocene and Holocene. Most of the erosion events were brief; that for the Lower Flamboura alluviation had a duration of a few centuries at most, and the most recent event, the Kranidhi, has accumulated since products made of plastic were introduced to the region. Between these short-lived events, soil profiles have formed which are indicative of prolonged slope stability. The timing of the first Holocene erosion episodes appears to follow some 500–1000 years after the introduction of agriculture. The southern Argolid was widely settled by farmers between the mid-fourth and mid-third millennia BC, but the Pikrodafni alluviation is not dated until the end of the third millennium. This suggests that it was not the introduction of agriculture *per se* which was responsible for soil erosion. Instead, erosion is likely to have occurred during periods of economic and probable political instability. 'Once the Greek landscape had been controlled by soil conservation measures, its equilibrium became precarious, the price of maintaining the equilibrium was high, and economic perturbations were only too likely to disturb it' (van Andel *et al.*, 1990a, p. 383). The characteristics, and in particular the degree of redness, of buried soil profiles or palaeosols, have allowed van Andel and his colleagues to recognise a sequence of soils which differ with maturity in several parts of Greece (van Andel, 1998) and which can help interpret the archaeological history of the region. The expulsion of the Moors from Spain and Portugal offers, to some writers, a parallel example of soil erosion occurring in times when terracing fell into neglect (McNeill, 1992; see Section 4.3.1). If damage to terraces caused by livestock is not kept in check, then gully erosion soon removes soil previously stored by the terraces. This is then laid down in the valley bottoms as streamflood deposits. This hypothesis is not accepted by all Mediterranean researchers.

8.11 A model of landscape change

For Mediterranean Europe, Grove and Rackham (1998) have produced an illustrative model of land use which is shown in Figure 8.7. While this cannot be considered representative of the whole Mediterranean Basin, it attempts to draw together some of the themes outlined above. The first time slice represents the prehistoric period. Sea levels were lower than at present, and some badlands and gullies had already developed. Mesolithic sites were located in caves and in coastal plains. From Neolithic times onwards, the landscape had become 'humanised'. Terraced and irrigated agriculture were being introduced and forest had declined; that which survived was managed and utilised, along with wetlands. By Roman times, new crops and technology had been introduced; carobs and mulberries from Asia and, the use of storage dams and aqueducts. Drainage of wetlands for arable land was taking place.

Not surprisingly, different regions experienced different population histories and landscape changes. There were phases of expansion and reduction, with associated decline and recovery of vegetation communities. Periods of peak population included the Hellenistic period (c. 200 BC) in Greece, the Imperial Roman period (c. 200 AD) and in Crete the Early Byzantine period

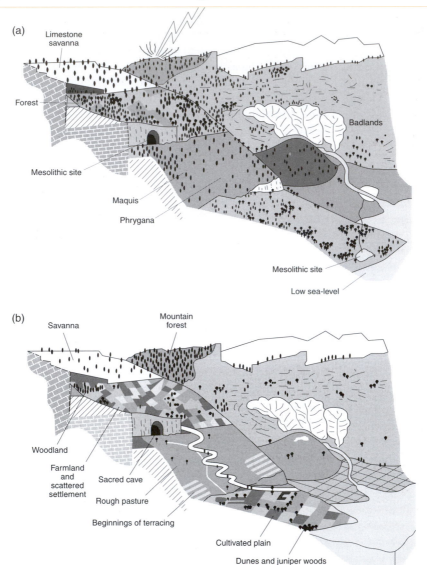

Figure 8.7   Model of land-use change for the northern Mediterranean Basin. After Grove and Rackham (1998). (a) BC, (b) AD 50, (c) AD 1770, (d) AD 1990.

(c. 560 AD) and Late Venetian period in the second half of the sixteenth century. In Sardinia, Sicily, Crete and most of the Iberian Peninsula, the expansion of the Arabs from North Africa introduced further new crops (sugar cane, rice, cotton, citrus fruits) and extended irrigation into terraced areas. There are indications of landscape stability at this time (McNeill, 1992) until the final expulsion of the Moors from Spain between 1570 and 1614, amount-

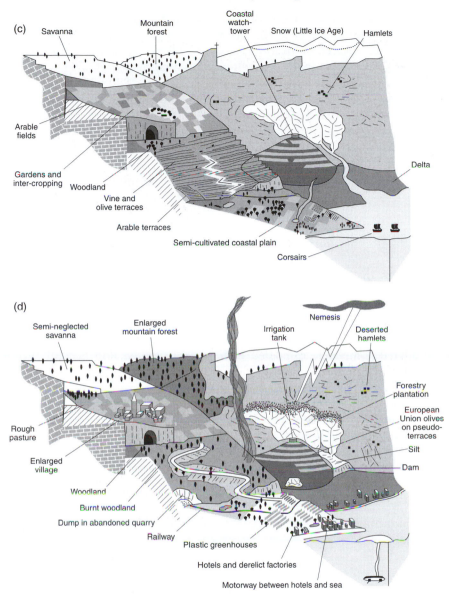

Figure 8.7 (*Continued*)

ing to a loss of a quarter of a million people. Population decline also often resulted from episodes of plague, such as repeated instances during the sixth to eighth, and fourteenth to seventeenth centuries. Plague could kill a third or more of local populations. In some areas along coasts and marshy rivers, malaria was a problem. Some coastal localities became depopulated as people retreated inland, especially in those subjected to piracy. This is shown in the time slice for AD 1770, a year typical of a time of droughts, floods, severe winters and famine associated with the Little Ice Age (Grove, 1988). Land use at the time showed a typical change with relief: forest and savanna on the higher slopes, terraced cultivation for vines, olives and arable crops on the middle to lower slopes, and semi-cultivated coastal plains.

Population numbers began to increase from about 1800 with people moving back towards the coast. Public land passed into private ownership and terraces were cut into hillsides, often to the limits of cultivation. The arrival of the railways meant that perishable crops, such as fruit, could now be grown for export to northern Europe. At the same time, the import of cheap grain meant the decline in local cereal cultivation. In general, southern Europe was not prosperous in the period 1830 through to 1950. Emigration became important, for both people and landscapes. Rural depopulation in southern Europe meant recovery of forest land (see Barbero et al., 1990) in upland areas, but expansion of urban areas. People also moved from southern Europe to industrialised northern Europe, North America and Australia. There was significant movement of people from France to North African colonies. In North Africa population expansion was occurring.

The final time slice (c. 1990) shows, for the northern Mediterranean, a largely abandoned agricultural landscape, replaced by one of industrial activity. Today, European Union policies dominate much of the rationale for recent landscape changes, especially those relating to agriculture. Coastal regions have altered completely compared with previous times. Post Second World War urbanisation and industrialisation, together with increases in mass tourism, have placed particular demands on water. Away from the narrow coastal plains, however, semi-natural vegetation has recovered, albeit with changed species composition. Many European colonists returned from North Africa during the period 1950–1990 and many North Africans emigrated to Europe, especially Spain, France and Italy. Population expansion continued with large-scale migration to urban areas but populations remained high in rural regions, putting significant pressures on the landscape.

8.12 Summary

Land use and land-use changes have had a considerable impact on Mediterranean landscapes. Some practices, such as grazing and the use of fire, have resulted in the development of vegetation and animal communities (with high species richness and rates of endemism) dependent on their continued use. Elsewhere modern agricultural intensification has raised fears of landscape instability – erosion and degradation – though neither of these is new to the region. The influences of climatic variability, tectonic activity, geological and topographic heterogeneity and different soil types lead to the potential for high rates of soil erosion. Demographic and social changes in the past fifty years have led to fears for the survival of some Mediterranean ecosystems. The resultant threats and their associated environmental issues are summarised in Table 8.2 and conservation measures to address some of these are the subject of the next, and last, chapter.

Table 8.2 Threats to Mediterranean habitats

Threat	Effect on habitat and/or animal populations	Habitat threatened			
		CF	BF	M	G
Abandonment of grazing, or undergrazing	Increases vegetation cover; reduces vegetation diversity; increases vegetation height; loss of open areas	×	×		×
Frequent and large fires	Opens up habitat and simplifies structure; increases susceptibility to erosion and, in extreme cases, 'desertification'	×	×	×	×
Lack of fire	Increases vegetation cover			×	×
Afforestation with native and non-native trees, e.g. eucalyptus	Habitat loss	×	×	×	×
Recreation, hunting	Disturbance to breeding and feeding of animals may lead to reduced adult survival, increased predation and starvation of young	×	×	×	×
Forestry					
Coppicing	Initially opens habitat and reduces structural diversity; eventually leads to dense growth		×		
Selective felling	Promotes regeneration, but reduces canopy and vertical structural diversity; removes tall emergent trees; reduces food resources and breeding sites	×	×		
Clear-felling	Initially opens up habitat, promotes erosion, and leads to uniform regrowth; often followed by afforestation	×	×		
Undergrowth removal and thinning	Destroys shrub layer, promotes fast growth and uniform structure	×	×		
Reservoir creation	Habitat loss	×	×	×	×
Wetland drainage	Habitat loss	Mostly of coastal and riparian sites			
Urbanisation	Habitat loss	×	×	×	×
Transport, infrastructure	Habitat loss	×	×	×	×
Industrialisation	Habitat loss; pollution	×	×	×	×
Conversion to perennial crops	Habitat loss			×	×
Conversion to arable	Habitat loss		×	×	×

Table 8.2   (*Continued*)

Threat	Effect on habitat and/or animal populations	Habitat threatened			
		CF	BF	M	G
Overgrazing	Promotes very short vegetation structure, bare ground, excessive erosion and high disturbance; prevents regeneration of woody species	×	×	×	×
Tree diseases	Change in vegetation structure; increase in dead wood	×	×		
Powerlines	Mortality of large bird species; if lines at sufficiently high density, habitat avoided	×	×	×	×
Habitat fragmentation	Reduces plot size, increases edge habitat, landscape diversity, habitat insularity, susceptibility to disturbance	×	×	×	×
Pesticide and herbicide use in forestry and agriculture	Loss of invertebrate food resources, disrupts ecosystems, potential mortality from toxic products	×	×	×	×
Mining, quarrying	Habitat loss	×	×	×	×
Bulb collection	Reduction in population numbers		×	×	
Introduced species		×	×	×	×

CF, Coniferous forest; BF, broadleaved forest; M, maquis; G, garrigue, rocky.
Sources: After Médail and Quézel (1997) and Tucker and Evans (1997).

8.13 Further reading

Grove, A.T. and Rackham, O. (2000) *The Nature of Mediterranean Europe: An Historical Ecology*, Yale University Press, New Haven, CT.
Mannion, A.M. (1997) *Global Environmental Change*, 2nd edition, Longman, Harlow.
Roberts, N. (1998) *The Holocene*, 2nd edition, Blackwell, Oxford.

Chapter 9

Environmental issues and conservation

In 1637 at an auction of bulbs in Alkmaar, the Netherlands, 180 tulips, narcissus, anemones, lilies and carnations were sold for 90 000 guilders, the equivalent of about £6 million today. The country was gripped by tulip-mania as breeders tried to produce ever more delicate and ornate varieties. Ultimately there were too many speculators and too few buyers and the market collapsed spectacularly within weeks of the auction. Tulip-mania had only lasted a short time – tulips were not introduced to western Europe until around 1560 and the first bulbs were not planted in the Netherlands until 1593 (Pavord, 1999). The bulbs were originally imported from Turkey, where they grow wild. Today, Turkey still exports flower bulbs, mostly to Dutch wholesalers who re-export to Germany and the United Kingdom. In the mid-1980s some 60 million bulbs were being collected from the wild each year, in particular species of cyclamen, anemone and snowdrop. Fears for the sustainability of the wild populations grew, compounded by reports from bulb collectors of smaller harvests and the need to collect from ever more inaccessible locations. Fortunately such fears have led to conservation initiatives developed with local people to try to stem the bulb exploitation. Villagers in the relevant areas are now discouraged from collecting from the wild for export, but are encouraged to plant some bulbs on village land to grow on, and then harvest as a sustainable crop each year. The initial results of the schemes appear to be successful (see Section 9.4.1) and may lead the way for other conservation partnerships for threatened species and habitats around the Mediterranean.

Threats to the Mediterranean landscapes and their associated environmental issues were examined in the previous chapter and summarised in Table 8.2. Once these have been identified conservation strategies can be designed and implemented. This does not just mean conservation of biological resources, such as individual species of plants and animals, but also conservation of physical resources, such as the soil. In addition any measures to conserve or protect biological and physical resources need to be set within the context of international and national initiatives. This is by no means easy to achieve; conservation biologists are not even unanimous as to the best ways in which to proceed – whether to aim for species or habitat or ecosystem or landscape conservation. There are also conflicting interests; protecting vegetation cover in the interests of reducing soil erosion might lead to exclusion of grazing on maquis-covered slopes. Yet exclusion of grazing could lead to reduced species richness and a greater risk of fire as inflammable litter builds up. Conservation initiatives need therefore to integrate the biological and physical components of the Mediterranean world. Conservation matters related to the environmental issues already highlighted in this book are the subject of this last chapter, beginning with those relevant to biological resources.

9.1 Conservation – the value of biodiversity in the Mediterranean context

Why should societies be interested in the conservation of wildlife? This question and any answers to it are inevitably a part of a philosophical debate about the relationship between people and their environments, but as problems of environmental change (whether anthropogenically induced or not) become more evident, so societies have increasingly realised the extent to which they rely on wildlife and ecosystem processes for their survival. This might be through the way in which vegetation cover inhibits soil erosion, or through the role that wetlands play in dampening the effects of flooding, or through the medicinal properties of some plants and animals. Whatever the recognition, it increasingly involves an economic realisation that plants and animals have a value (Edwards and Abivardi, 1998). Sometimes these are use values, for example the direct economic value that comes from exploitation of fish and timber or from using a landscape for recreation and tourism. Some use values may derive from the ecological function provided by ecosystems – the benefits of flood control, nutrient cycling, photosynthesis and the primary productivity of plants. Some benefits are regarded as option values, such as future foods or genes for plant breeding. In other words conservation of species gives us options for the future. Other values may be regarded as non-use values, for example the benefits that come from the satisfaction that wildlife and ecosystems exist and the desire to pass these benefits to future generations. Underlying this recognition of the value of wildlife lies the fear that biodiversity is a diminishing resource and that environmental degradation is occurring at an ever increasing pace. The value of biodiversity has been placed at the centre of conservation debates by the United Nations Convention on Biological Diversity (see Box 9.1). However, in spite, and maybe even because, of the central role of biodiversity there remains the problem of how best to initiate conservation and continue with those projects already under way. This is the subject matter of the next section with reference to the Mediterranean region.

9.1.1 Mediterranean biodiversity loss

The ultimate loss of biodiversity is the extinction of species, but as extinction is a common and natural component of evolution, what concerns society today is that the rate of extinction has been increasing as human activity has intensified. At a global scale, there is evidence to suggest that the rates of extinction of vertebrates and vascular plants are now 50 to 100 times the

Box 9.1   Institutional frameworks for conservation

The impetus for conservation in many Mediterranean countries is provided by a number of international conventions. These provide a strong basis for co-operation between nations and impart a strong obligation to comply with the conventions' directives. However, the implementation of these directives can only be as good as the national legislation of each signatory country and cannot be taken for granted. In part this stems from the fact that the wording of some conventions is sufficiently vague to allow for political expediency in trying to get the conventions accepted by as many nations as possible. The conventions described below are those applicable to Mediterranean environmental issues and nations. However they do not each apply to every nation within the region.

The *Convention on Biological Diversity*, adopted following the United Nations Conference on Environment and Development (the 'Earth Summit' held in Rio de Janeiro in June 1992) places a firm obligation on each treaty nation to ensure the conservation of its biological diversity. Governments have to prepare national strategies, plans or programmes for the conservation and sustainable use of a nation's biological resources and integrate these into relevant economic, social and political policies. These strategies should promote the protection of ecosystems and natural habitats, and the maintenance of viable populations of species in their natural surroundings and, where necessary, restore degraded ecosystems and promote the recovery of threatened species. This convention is probably the most important broad legislation for maintaining biodiversity (Tucker and Evans, 1997) as it sets the framework for each nation's own strategy.

The *Convention on the Conservation of Migratory Species of Wild Animals* or the 'Bonn Convention' is another which has worldwide coverage. Its objectives are the conservation and effective management of terrestrial, marine and avian species over the whole of their migratory ranges. Part of this convention includes an agreement on African–Eurasian migratory waterbirds and is therefore of particular relevance to the Mediterranean region.

Countries which are signatory to the *Ramsar Convention on Wetlands of International Importance especially as Waterfowl Habitat* aim to designate and promote the conservation and wise use (sustainable utilisation) of suitable wetlands within their territories. As such this is the most important international legislation for the protection of wetlands – not just coastal but also inland wetlands. The convention introduced the widely used concept of '1% flyway population levels' for overwintering waterbirds when assessing the importance of sites. In the Mediterranean region there are some 130 Ramsar sites.

Other conventions which cover the Mediterranean region include the *United Nations Convention on the Law of the Sea* and the Barcelona *Convention for the*

(continued)

(continued)

Protection of the Mediterranean Sea against Pollution, also known as the 'Mediterranean Action Plan' for the protection of the Mediterranean environment and the sustainable development of its coastal areas. The *Convention Concerning the Protection of the World Cultural and Natural Heritage* aims to ensure the protection of natural and cultural areas of outstanding universal value. In the Mediterranean region there are 15 World Heritage Sites.

The European nations of the Mediterranean are bound by several pan-European directives, in particular the *Convention on the Conservation of European Wildlife and Natural Habitats*, also known as the Bern Convention. Its member states are the 38 countries of the Council of Europe (1996 data) and invited non-member states in Europe and North and West Africa, as well as the EU itself. It was negotiated in the late 1970s to conserve wild flora and fauna and their natural habitats, especially those species whose conservation requires the co-operation of several nations. The need to implement the convention was the motivation for the European Union *Habitats Directive* (1992). This obliges all EU Member States to promote the maintenance of biodiversity, taking account of economic, social and regional requirements, through measures to restore, at favourable conservation status, natural habitats and species of wild fauna and flora of Community interest in the European territory of the Member States. A minimum standard for biodiversity conservation is set by creating a coherent ecological network (Natura 2000) of Special Areas for Conservation (SACs), identifying priority habitats and species and monitoring their conservation status. To protect SACs and the overall coherence of Natura 2000, assessments of adverse impacts must be made and measures taken to mitigate any projects which have to be carried out at or near a SAC for imperative reasons of public interest. The Habitats Directive has the potential to protect a wide range of species throughout the Union but its success depends on the individual commitments of the Member States.

An earlier Directive (1979) which also applies to EU Member States is the *Birds Directive*, which covers the conservation of all species of naturally occurring birds in the wild state in the European territory of the Member States, covering their protection, management and control. It also lays down rules concerning their exploitation. Other EU Directives which involve elements of conservation and environmental protection include the *Nitrates Directive* to reduce water pollution caused by nitrates from agricultural sources, the *Directive on the Conservation of Fresh Water for Fish*, the *Directive on Urban Waste Water Treatment* and the *Directive on the Freedom of Access to Information on the Environment*.

Conservation costs money and financial help is available to nations from several sources. The *LIFE* fund of the European Union is intended to assist in the development and implementation of the Union's environmental policies. Reform of the EU's Common Agricultural Policy established the

(continued)

(continued)

Agri-environment Regulation which aims to protect the farmed environment through payments to farmers who follow environmentally beneficial farming practices which protect and manage the countryside. As such farmers are compensated for reducing production and aiding the environment.

The *Convention on the International Trade in Endangered Species* (CITES) is also of relevance. It regulates a worldwide system of controls on international trade in threatened wildlife, which are placed in two main categories. Appendix I includes all species threatened with extinction which are or may be threatened by trade. Appendix II includes all species which although not necessarily currently threatened with extinction may become so unless trade is subject to strict regulation. It also applies to certain species which are similar in appearance to the previous group so as to avoid mistaken identity in trading. In addition to the two main categories, Appendix III species include those which any CITES signatory country deems subject to regulation within its own jurisdiction for the purpose of preventing or restricting exploitation. Government permits are required for all trade in listed species. Bulbs of the genera *Cyclamen* and *Galanthus*, both traded for years by Turkey, are listed in Appendix II. Turkey's implementation of CITES came into force in December 1996 and the country now sets quotas for the harvest and export of these bulbs (see Section 9.4).

Source: Tucker and Evans (1997).

---

expected background rates (Heywood and Watson, 1995). Extinctions of large mammals in the Mediterranean region during the last three millennia, for example the lion (*Panthera leo*) and the wild ass (*Equus asinus*), have already been discussed in Section 6.4, but the most recent extinction appears to be a subspecies of the Pyrenean mountain goat (*Capra pyrenaica pyrenaica*). The last known specimen was found dead in January 2000 (*The Guardian*, 11 January 2000). Extinctions have also occurred in the plant world: 33 taxa (31 species and 2 subspecies) are listed as extinct in the region (Greuter, 1994). This compares with a total of about 29 000 taxa (some 23 000 species plus subspecies) described as native to the Mediterranean. This is an extinction rate of 0.13% at the species level, or 0.11% of the taxa. Obvious questions are whether the data are reliable enough to make such a calculation and whether such a rate should be a cause of concern.

Around the Mediterranean Basin few plant species are so well studied and monitored that their every locality is known. It is therefore likely that more species are extinct than are known to be. Some regions such as the Maghreb, with a high level of endemism, report disproportionately low levels of extinction compared with other parts of the basin (Greuter, 1994). By contrast some species thought extinct might actually survive in unknown locations. In Greece three species presumed missing have been re-recorded. The only known locality

Table 9.1   IUCN Red List categories

The Categories

Extinct (EX)	A taxon is Extinct when there is no reasonable doubt that the last individual has died.
Extinct in the Wild (EW)	A taxon is Extinct in the wild when it is known only to survive in cultivation, in captivity or as a naturalised population (or populations) well outside the past range. A taxon is presumed extinct in the wild when exhaustive surveys in known and/or expected habitat, at appropriate times (diurnal, seasonal, annual), throughout its historic range have failed to record an individual. Surveys should be over a time frame appropriate to the taxon's life cycle and life form.
Critically Endangered (CR)	A taxon is Critically Endangered when it is facing an extremely high risk of extinction in the wild in the immediate future.
Endangered (EN)	A taxon is Endangered when it is not Critically Endangered but is facing a very high risk of extinction in the wild in the near future.
Vulnerable (VU)	A taxon is Vulnerable when it is not Critically Endangered or Endangered but is facing a high risk of extinction in the wild in the medium-term future.
Lower Risk (LR)	A taxon is Lower Risk when it has been evaluated, does not satisfy the criteria for any of the categories Critically Endangered, Endangered or Vulnerable. Taxa included in the Lower Risk category can be separated into three subcategories: (1) *Conservation Dependent* Taxa which are the focus of a continuing taxon-specific or habitat-specific conservation programme targeted towards the taxon in question, the cessation of which would result in the taxon qualifying for one of the threatened categories above within a period of five years; (2) *Near Threatened* Taxa which do not qualify for Conservation Dependent, but which are close to qualifying for Vulnerable; (3) *Least Concern* Taxa which do not qualify for Conservation Dependent or Near Threatened.
Data Deficient (DD)	A taxon is Data Deficient when there is inadequate information to make a direct, or indirect, assessment of its risk of extinction based on its distribution and/or population status. A taxon in this category may be well studied, and its biology well known, but appropriate data on abundance and/or distribution is lacking. Data Deficient is therefore not a category of threat or Lower Risk. Listing of taxa in this category indicates that more information is required and acknowledges the possibility that future research will show that threatened classification is appropriate. It is important to make positive use of whatever data are available. In many cases great care should be exercised in choosing between DD and threatened status. If the range of a taxon is suspected to be relatively circumscribed, if a considerable period of time has elapsed since the last record of the taxon, threatened status may well be justified.
Not Evaluated (NE)	A taxon is Not Evaluated when it is has not yet been assessed against the criteria.

Source: *IUCN Red List Categories*, IUCN Species Survival Commission (1994).

of *Onobrychis aliacmonia* (a member of the pea family) was flooded in the
1970s with the construction of a dam on the Aliakmon river, but in 1985 it
was rediscovered some 400 km away in the Peloponnese. Ironically here too
the number of individuals is declining and it may become extinct once
more, probably not to be discovered a third time (Greuter, 1994). It is also
likely that an unknown number of species became extinct before botanists
first began to study and describe the flora. This means that the current rate
of 0.11% may be misleading in the longer term. Other regions with a
mediterranean-type climate, which have a shorter history of land use, have
higher current rates of extinction, for example 3% in the fynbos biome in
South Africa and possibly as high as 6% in Western Australia (Greuter, 1994).
It is also important to recognise that as well as species which are already
extinct many more may be threatened with extinction. Therefore an extinction
rate of (only) 0.11% means relatively little when considered in isolation.

Monitoring the extinction threat of plants and animals is the role of the
International Union for the Conservation of Nature and Natural Resources. It
compiles Red Lists or red data books of threatened species, placing them into
a number of risk categories (Table 9.1). Red data lists of plants and animals
are produced country by country, which makes it very difficult to assess the
proportions in the different categories for the entire Mediterranean region;
some species will be listed under more than one country, some will have
a range outside the Mediterranean part of that country. Nevertheless, of the
2879 species identified as endemic to individual Mediterranean countries
(excluding Syria, Lebanon and Turkey), 1529 were believed to be rare or
threatened in the 1980s according to categories pertaining at that time (Leon
et al., 1985). Of the remaining endemics there was insufficient information to
categorise nearly 300 of them. Unfortunately being listed in a red data book,
whether for plants or animals, does not offer any guarantee of protection and
survival. For that, targeted conservation programmes are needed. The data lists
do however provide benchmarks for future assessments.

9.2 Conservation – the value of the physical landscape and its integration with the biological resources of the Mediterranean region

The MEDALUS project (see Chapter 4) has the objective of understanding
the complexities of land degradation in the European Mediterranean. This is
the reduction and loss of economic and biological productivity resulting from
land uses or from a process or combination of processes arising from human
activities, such as soil erosion, deterioration of the physical, chemical, biological and economic properties of the soils and long-term loss of vegetation
(Thornes, 1998a). As such, the project seeks to integrate the biological, physical and economic facets of the region, but central to this is an appreciation of
the role played by soils. The ability to compare physical and biological processes across the Mediterranean Basin provides a powerful tool for assessing
the extent of the issue – whether 'desertification' exists as a problem and

Plate 9.1   Villa development for second homes encroaching on previously agricultural land.

Plate 9.2   Intensive agriculture. Lettuce grown in a Mediterranean polytunnel. © David Woodfall/Woodfall Wild Images.

whether it is site specific or ubiquitous – and for devising appropriate land-use conservation measures.

Undeniably, several socio-economic factors have led to an emphasis on land-use conflict and its influence on soils and soil erosion, particularly in the past fifty years: improved living standards, the development and spread of tourism (Plate 9.1), agricultural intensification (Plate 9.2) but also abandonment of some traditional terraced cultivation, rural depopulation and a shift to urban living. Ultimately all these have increased demands for scarce water. It is therefore not surprising that land degradation or desertification is perceived as a problem. However, early indications from the MEDALUS results suggest that the extent of the problem may not be as great as once feared. Initial results of measured soil erosion rates at field plots are generally significantly lower than commonly supposed (Thornes, 1998b). These results may have occurred because measurements were made during a run of generally very dry years and, indeed, it may not be total annual rainfall that is the determining factor but the occurrence and frequency of intense rainstorms. We also need to be cautious when scaling results from field plots to catchment areas. Therefore we need to avoid being over-optimistic in extrapolating rates of soil loss across the basin and into the future. Nevertheless, to reduce runoff and erosion, effective vegetation cover is needed and all the MEDALUS work supports this idea. To this end the encroachment of woodland onto old terraced land can be regarded as a positive landscape change, even though it occurs with an increased risk of fire. Recovery of vegetation will also aid recovery of soils – MEDALUS model simulations suggest progressive increases in soil organic matter content, improved soil structure and increased infiltration rates. In the short term re-vegetation will not solve the problem of water shortage as over-abstraction has lowered groundwater tables. However, it is suggested that in the longer term, and carried out in an integrated fashion at the scale of the watershed, restoration of vegetation and soils will reduce flood risks, decreasing soil erosion and promoting infiltration; careful water management will always be needed, especially in a future with uncertain changes in precipitation.

A consequence of reduced rainfall, higher temperatures and increased vegetation cover may be a greater incidence of wildfires, whether natural or anthropogenically induced. Fire monitoring is therefore essential. Multinational projects in the Mediterranean region to assess environmental issues like this are becoming increasingly common. The MEGAFiReS project uses remote sensing and seeks to detect forest fires in the Mediterranean region by automatically searching for fire-saturated pixels on sensed images. Various aspects of fires can be sensed, such as the energy released by a fire, its smoke, char and fire scars (Robinson, 1991). Although designed for meteorological observations, the AVHRR (Advanced Very High Resolution Radiometer) on the NOAA polar orbital satellite has been widely used in fire detection studies and also in assessments of vegetation water stress and therefore risk of fire. NOAA-AVHRR data have good temporal resolution, with two images per day, and good spectral information with four channels sensing visible, near, mid- and thermal infrared (IR) radiation. Under the MEGAFiReS project daily images are collected and their spectral signature analysed (Chuvieco et al., 1999). Pixels containing radiometric values characteristic of active forest fires and burnt surfaces can be identified. Coverage of water bodies and urban areas where forest fires cannot occur are disregarded. There are, however, some problems in the use of mid-IR data, particularly its sensitivity. It was originally designed to estimate sea surface temperatures and so it becomes saturated at 320 K, therefore confusions between fires and other hot surfaces (such as bare soil in summer months) can easily occur. In addition some pixels may appear to contain burnt land, even if only a small proportion of it is actually occupied

by a fire, because there is such a strong contrast between a fire and its surroundings. Some fires will also be missed because they are of such short
duration that they are no longer active at the time of satellite overpass. Cloud
cover further interferes with sensing. Continuing research in the MEGAFiReS
project intends to refine analysis of burning and the prediction of fires. This
will enable a better response to burning when it occurs.

Fire scientists have created fuel models which summarise vegetation structural properties into a few vegetation types, which can then be identified and
modelled. In addition, remote sensing can determine the fuel moisture content of vegetation (Chuvieco *et al.*, 1999). This is obviously important because
the drying of both living and dead fuel increases the risk of fire. Fuel moisture
content is determined by plant reflectance and emittance, which can be sensed.
Remotely sensed data have to be validated by ground surveys. In southern
France fuel moisture content is measured twice weekly during the summer
months at a variety of sites. Measurements are made on typical maquis species
such as *Quercus ilex*, *Q. coccifera*, *Cistus monspeliensis*, *C. albidus*, *Erica arborea*,
Rosmarinus officinalis and *Juniperus oxycedrus*. The data are passed to local and
regional fire-fighting authorities but have also shown the potential for using
NOAA-AVHRR-derived data for monitoring fuel moisture content. The use
of such data should improve estimates of short-term fire risk.

Fire ecologists have recognised for some time, whereas politicians may not
have accepted, that fire suppression is costly and often futile. This follows the
recognition that conservation management strategies should no longer be based
on 'illusory succession-to-climax theories of non-interference' (Naveh, 1994,
p. 178). Instead it is necessary to re-establish homeorhetic flow equilibrium
(Section 7.4) by simulating disturbance regimes. Prescribed fires can therefore
be used as a management tool. Fire has a long association with pastoralists
around the Mediterranean, who use it to create new grazing land. It checks
the growth of unpalatable vegetation and promotes lush regrowth of palatable species. Prescribed burning is also useful in managing fuel accumulation,
particularly in monotonous and highly inflammable conifer and eucalyptus
plantations.

9.3 Conservation in practice

9.3.1 Hotspots conservation

Conservationists are unable to assist all landscapes and species threatened by
human activity if only because of the costs involved. Inevitably this results in
a prioritising of threatened taxa and places. This underlines, for example, the
designations used by the International Union for the Conservation of Nature
and Natural Resources (IUCN) for threatened plants and animals and the
designation of global hotspots of biodiversity, of which the Mediterranean is
one. Some conservation planners argue that funding should be focused on
these areas; if some of the estimates of global species extinction that have been
proposed turn out to be valid, then between one-third and two-thirds of all

species might be likely to disappear in the near future, but much of this could be reduced if the world's 25 hotspots were protected by means of a 'hotspots rescue fund' (Myers *et al.*, 2000). Of course the estimates of current and future extinction rates may well be very wrong.

The hotspots argument involves the recognition of species richness and high rates of endemism as valuable measurements of ecological diversity, indicating where there is a need for conservation strategies. Endemics may be restricted in their geographic range or in their ecological range, or both. Geographic endemics are those found only within a geographical locality, such as Spain or Portugal, whereas ecological endemics are those restricted to a habitat-type, for example maquis or garrigue. The distinction needs to be made because not all endemics are equally rare or threatened with extinction. This is illustrated by the heathlands of southern Spain where plant species richness is generally low, but endemism high (Ojeda *et al.*, 1995). The endemic species are also those most abundantly occurring within their preferred habitats. It is suggested that the acid, nutrient-poor soils around the Gibraltar Strait may have been an important factor promoting speciation and endemism, as the adverse soil conditions led to the dominance of the tolerant species. The relative abundance of the endemics means that they are not apparently endangered. However, as well as species richness and endemism, Ojeda *et al.* (1995) measured a third component of biodiversity – taxonomic singularity. This is a measure of the number of species within each genus. This parameter provides some clues to the evolutionary history of the genus but cannot be taken to reflect genealogy directly. However, broadly speaking a species within a less diverse genus (i.e. a genus with relatively few species) is more distinct than a species belonging to a genus containing many species. This means that not all species can be considered as equal in value in terms of their evolutionary history. In the acid heathlands where endemism is high, taxonomic singularity is low because species come from very diversified genera such as *Erica* (heath), *Cistus*, *Quercus* (oak) and *Genista* (broom). In contrast limestone shrublands show high taxonomic singularity, in other words non-diversified genera such as the single-species genera *Olea* (olive – *Olea europaea*), *Myrtus* (myrtle – *Myrtus communis*) and *Chamaerops* (dwarf palm – *Chamaerops humilis*). As these species are very frequent throughout their range (circum-Mediterranean) they are not in need of special conservation status. There might, however, be instances when taxonomic singularity correlates with rarity and adds to the conservation value of a habitat. This is possibly the case with semi-deciduous oak woodlands in the region which have high singularity, but also have a large number of subspecies present which have low abundance. Ojeda *et al.* (1995) make a compelling argument that assessment for conservation needs to be based on more than just rarity and endemism.

Outright protection of areas with high species richness is not a feasible option, especially where some of that diversity is a consequence of vegetation community development alongside millennia of human land use, as in the Mediterranean Basin. In addition, there are flaws inherent in the hotspots conservation argument – we cannot really know how species rich an area is, especially if it is one that is under-explored. Many hotspots are merely those areas where people have gone collecting. Efforts need to be expended in non-hotspot regions, in other words the conservation of all types of ecosystems, not just those found within hotspots.

Conservation with a priority for making inventories of, and rescuing, rare and endangered species has been characterised as the 'fire-brigade period' of conservation (Edwards and Abivardi, 1998) and while conservation of rare and endangered species will remain a goal for many ecologists, there is recognition of the need to move on from this approach. Moving on entails a shift from protecting small nature reserves and their plants and animals to larger landscapes. This broadens the spatial scale from the species and ecosystems level to that of semi-natural and cultural landscapes – in the Mediterranean region at least, there are very few, if any, natural landscapes remaining. At the same time the shift in emphasis should incorporate the economic value of biodiversity and place conservation firmly within any nation's plans for its economic development.

9.3.2 Landscape conservation

Just as the IUCN produces red data lists of threatened plant and animal species, so it is argued that there should be analogous Red Books of threatened landscapes (Naveh, 1993). These would involve the identification of the geographical, ecological and cultural components of a landscape. This scale is crucial because most decisions about land uses, which determine the fate of wildlife and their habitats, take place at the regional level of the landscape. 'Therefore threats to such tangible landscapes and their biodiversity and environmental quality may have much more meaning and appeal to the public than threats to species or vaguely defined ecosystems' (Naveh, 1993, p. 245). At this scale practical measures for conservation of the physical landscape, such as combating soil erosion, should be more possible. Landscape Red Books (later renamed Green Books) should suggest practical solutions to prevent further environmental degradation, serve as guidelines for political decision making and provide an opportunity to consider realistic conservation in the context of the socio-economic and cultural characteristics of the threatened region. While appearing idealistic, the concept is being used to evaluate the threats to the landscape of western Crete (Grove and Rackham, 1993; Papanastasis and Kazaklis, 1998). Other landscape-scale conservation schemes include those aimed specifically at regions of geological interest such as the Puntos de Interés Geológicos, or Points of Geological Interest, in Spain (Alcalá, 1999). These are areas characteristic or representative of a region's geological heritage with sites classified according to their stratigraphic or palaeontological content.

9.3.3 Protected areas in the Mediterranean

Conservation and environmental protection are regarded as important national considerations to the countries of the Mediterranean. With the exception of

Malta and Syria all have sites designated as protected areas according to United Nations criteria, and Malta has protection measures to safeguard its environment. In addition most countries have national strategies for conservation. Protected areas such as national parks and forests, nature reserves and nature monuments therefore play an important part in current conservation initiatives around the Mediterranean and they will continue to do so with the implementation of the European Union Habitats Directive to create an ecological network (Natura 2000) of Special Areas for Conservation (SACs) (see Box 9.1). Assessing the success of these in Mediterranean countries will be difficult because the quality of protection and legislation to enforce this varies from nation to nation, but there are some 400 protected areas as designated by the United Nations. Many of these are for multiple use – not just the protection of wildlife but for recreation (for example Coto Doñana in Spain), sometimes for hunting and often for agriculture; only a small proportion will be free of some kind of human use. Indeed suppressing human activity may be to the detriment of wildlife. As outlined in Chapter 7, some Mediterranean landscapes are perturbation-dependent and therefore the prevailing land-use practices need to be maintained. A good example comes from La Encantada botanical reserve in Albacete province, southeast Spain. At present the reserve is grazed by sheep. They prevent the shrub cover from encroaching into the areas between the rosemary (*Rosamarinus* spp.) stands. If grazing ceased the shrubby vegetation would probably rapidly develop and eliminate the shade-intolerant weedy endemics. This latter group colonises the disturbed ground which comes with grazing (Gómez-Campo and Herranz-Sanz, 1993).

Designation of further protected areas will need to take account of global change, such as a warmer climate. The distribution of species and ecosystems will change, though not *en bloc* because of variations in dispersal rates of different organisms and because some may not survive environmental change. Therefore tomorrow's protected areas will not contain today's ecosystems (Holdgate, 1994). In consequence plans and policies need to accommodate these changing patterns so as to maintain as much biodiversity and wildlife habitat as possible, even in places where human land uses intensify in the future. This is part of the challenge facing the European Union Habitats Directive, and there are already some indications, from an analysis of its impact in Spain, that it does not take sufficient account of the degree of threat to individual taxa (Domínguez Lozano *et al.*, 1996). A comparison of the threatened flora in the Directive's Annex II has a correspondence of only about 60% with those classified as endangered by the Bern Convention (see Box 9.1), yet the two lists were published within months of each other. Part of the reason lies in the fact that the EU Directive is based on information provided by each Member State. As political boundaries rarely coincide with ecological ones, a species may be considered threatened in one country but not in another. For example the geophyte, *Narcissus asturiensis* (a species of daffodil) is found at a single locality in Portugal and so is listed as threatened, while in northern Spain it is widely distributed and generally classified as non-threatened. The problem of including taxa which do not require special conservation is compounded by the omission of others that do. For Spain 173 species regarded as endangered or vulnerable according to IUCN criteria are not included in the Habitats Directive. Under the terms of the Directive these would not be considered in need of Special Area Conservation.

An inherent problem with the use of protected areas as a basis for conservation strategies is the focus on their biological wealth. Measures aimed at protecting landscapes from soil erosion by promoting vegetation cover need to be implemented across the Mediterranean Basin and not just in nature reserves. In particular, the impact of intensive agriculture needs to be addressed. Within the European Mediterranean support for this approach comes from reform of the Common Agricultural Policy, through the 'Agri-environment Regulation'. This aims to protect the rural landscape through compensatory payments to farmers who follow environmentally beneficial practices. In 1997 some 17% of the total farmed area in the EU was subject to some form of management agreement under this regulation (Tucker and Evans, 1997). The environmental practices followed by farmers are more likely to benefit the more dispersed species of wildlife which exist on farms rather than the most endangered species. Environmental measures in EU Member States may also be helped by the Structural Funds of the EU. Their main purpose is to achieve economic and social cohesion within the Union, in particular to encourage development in those regions which have a GDP less than 75% of the Community average. This has applied to much of the Mediterranean regions of Italy, Spain, Portugal and Greece. In addition there are two further objectives relevant to the environment – those related to the adjustment of agricultural structures and concerned with the wider aspects of rural development.

There is, however, a problem in integrating development and environment, sadly rarely an achievable outcome. Many of the environmental problems facing the Mediterranean region today appear to stem from urbanisation and agricultural intensification since the Second World War. It would therefore be all too easy to argue for a reversion to the 'traditional' land-use practices of an earlier period, some of which still exist in the peasant economy of the region, in the belief that these were in equilibrium with the environment. This is, of course, disingenuous and over-romantic. Rural Mediterranean life was, and still is, harsh and demanding, characterised by poverty, short life expectancy and a low level of productivity. Furthermore, much of the unpredictability of Mediterranean life flowed, and still does, from the climatic variability and tectonic hazards of the region. Episodes of landscape change, such as high rates of soil erosion, are not just a product of modern agricultural methods. Therefore integrating social and economic development with environmental protection needs to be aided by the requirement for strategic environmental assessment and environmental impact assessment for all development plans, policies and programmes. At present legislation is still needed to define standards for these, as well as increased funding, perhaps provided by prospective developers (Tucker and Evans, 1997).

Arguably, for conservation purposes, European Union agricultural policies need to be redesigned with a shift from price support mechanisms to those supporting sustainable and environment-based agriculture – a redesign of farming systems which recognises that farmers do not just produce food, but also landscapes, biodiversity and rural economies, societies and communities. In other words, there is multifunctionality of agriculture.

9.3.4 Ecosystem conservation

Many different examples could be used to illustrate examples of practical conservation at the scale of ecosystems, but only two are given here – managed woodlands and wetlands. They have been chosen to reflect some of the repeated themes of this book, notably because of their disturbance-dependent nature.

9.3.4a *Dehesa* and *montado* woodland

Management of woodland has a long history in many regions of the Mediterranean (see Chapter 8) but woodland existence today is particularly threatened by abandonment of traditional practices, such as the *dehesa* and *montado* systems of Spain and Portugal. While traditional management systems ensure a sustainable use, their typical slow growth cycles put pressure on their owners to obtain positive and short-term financial returns on their investments. Most *dehesas* (some 82% or 2 800 000 ha) belong to large private estates. Traditional management allows for multiple use – grazing of livestock (which contributed 72% of total *dehesa* production in Spain in 1986), collection of edible fruit and nuts, firewood and cork, and hunting of game. Important herds of Iberian pigs, retinta cattle and merino sheep are found in *dehesas*. These breeds are able to survive in environmentally difficult conditions, but in some cases their survival has been jeopardised by the introduction of new breeds. Between 1955 and 1974 the population of Iberian pigs declined dramatically with the spread of African swine fever. The number of sows in Spain fell from 567 000 to 77 000. Although the population has largely recovered, in Portugal they were replaced by 'improved' breeds which did not flourish in *montados* and are not regarded as having the same flavour as the Iberian pig (Grove and Rackham, 2000).

In addition to their economic value, *dehesas* provide environmental resources. They are habitats for a variety of endemic species of flora and fauna, some of which are in danger of extinction, such as the Spanish imperial eagle (*Aquila adalberti*) and the lynx (*Lynx pardinus*). Spanish *dehesas* are also directly connected to African savannas and northern European forests by overwintering migratory birds. Rare resident birds are also to be found, such as the black vulture (*Aegypius monachus*) and the black stork (*Ciconia nigra*). In addition to their biodiversity *dehesas* provide various environmental functions such as delaying storm flood waters, helping to maintain groundwater levels, reducing soil erosion and protecting soil nutrient cycles. Yet *dehesa* woodlands are deteriorating in quality; trees are ageing and there is relatively little natural regeneration. The introduction or reintroduction of game species such as red deer, fallow deer and, to some extent, moufflon hinder the natural regeneration of woodland and also have a negative impact on roe deer, one of the most valued indigenous animals. The survival of traditional agri-silvo-pastoral land uses also plays a social role in maintaining rural communities. Yet *dehesas* are threatened by changing land-use practices, some of which are driven by changing international economic activities. Reduced demand for cork, particularly by the wine industry, is threatening the survival of the Portuguese *montados* of the Alentejo province as it becomes less economic to tend and harvest the cork oaks; agricultural policies of the European Union also encourage farmers to change or give up. The result will possibly be decline in a rural economy and its ecosystems. To that end, some areas have been designated as natural parks.

The national parks of Doñana and Tablas de Daimel include *dehesa* lands as do five large natural parks and several smaller ones in Spain. An inventory for conservation purposes at Cabañeros natural park, in central Spain, revealed 24 plant and 3 butterfly species endemic to the Iberian Peninsula plus populations of black vultures, imperial eagles and large mammals (Abbott *et al.*, 1992). In the early 1990s the conservation objectives of the park were concerned with the two birds and management of the park aimed to encourage their prey species, such as rabbit and partridges. Expansion of *Cistus* scrub will provide good habitat for these animals and for deer and boar – the main carrion for the black vultures. While favouring the large raptors, the expansion of *Cistus* has the added bonus of providing habitat for other birds, butterflies and flora. Surveys indicate high levels of endemic plants and high bird diversity in *Cistus* and *Quercus* scrubland. Conservation for the raptors makes these umbrella species for other taxa. Ironically, the wild boar (*Sus scrofa*), a popular game animal, could also be regarded as an umbrella species, for during the 1990s some *dehesas* were being acquired by urban absentee landlords for their hunting rights (Grove and Rackham, 2000). Continued traditional management of *dehesas* for hunting will benefit their other economic, social and environmental characteristics. Threats from increased wheat cultivation are also receding as it becomes clear that precipitation is too variable for successful production without extensive irrigation.

9.3.4b Wetlands

The loss and degradation of wetlands is rooted in social, economic and political processes but is often directly manifested through agricultural intensification and tourism (Hollis, 1992). Over the past few thousand years there has been a massive decline in Mediterranean wetlands – in Roman times 10% of Italy (some 3 million ha) was wetland; by 1990 this had diminished to only 190 000 ha. Similar proportions can be quoted for other countries. The main causes can be relatively easily identified: population expansion with its demand for higher living standards and demand for water; centralised planning procedures divorced from local conditions (for example, proposals to build a dam on the Ramsar site of Lake Tonga in Algeria); financial policies which encourage the drainage of wetlands for the benefits of farmers (70% of the cost of drainage in France during the 1970s was met by EU agricultural support grants); the problems of land tenure in wetland management (for example, in the Goksu Delta in Turkey, 80% of the land in the 1980s was publicly owned

Plate 9.3   Birds congregating on a *salina* in the Coto Doñana National Park wetland, Spain.
© Pascal Pernot/Still Pictures.

but privately farmed, leading to difficult management decisions); development of tourism (especially along the coasts of Portugal, Spain, Cyprus and Turkey); and, last but not least, the public perception of wetlands as wastelands of no economic value and of conservation of wetlands as *a priori* anti-development. However, this perception is changing with the realisation that wetlands have both use and non-use values (see Section 9.1).

Many wetlands have a long tradition of multiple use. In the Camargue, evaporation of seawater to produce salt is an ancient industry, but one which has grown steadily since the end of the nineteenth century (Britton and Johnson, 1987). Seawater is pumped into ponds or salinas and is progressively concentrated by evaporation in a hundred or so lagoons. These are arranged in a series of anastomosing circuits which allow different flow rates and the precipitation of different salts. Few activities are allowed within the salinas but in some areas there is limited hunting and fishing. Although salinas are probably among the most inhospitable coastal ecosystems they provide valuable habitat for a range of birds, plants and aquatic vertebrates and invertebrates, in particular breeding grounds for several species of gulls and terns and the flamingo (*Phoenicopterus ruber*). In winter significant numbers of migratory species, especially waders, congregate at Mediterranean salinas (Plate 9.3) (Tucker and Evans, 1997). Exploitation of the salinas is in line with 'wise use' of wetlands allowed under the Ramsar Convention (see Box 9.1) – sustainable utilisation for the benefit of people compatible with the maintenance of the natural ecosystem.

The recognition of wise use and the non-economic value of wetlands should aid in their conservation together with international legislation such as the Ramsar Convention and the European Union Birds Directive. Policies developed at the level of government should benefit from those developed by wetlands ecologists such as the Grado declaration and strategy 'to stop and reverse the loss and degradation of Mediterranean wetlands' (Finlayson *et al.*,

1992, p. 8). Unfortunately pressures on coastal wetlands are likely to increase. Numbers of tourists to the Mediterranean coastline are predicted to increase rapidly to between 380 and 760 million arrivals per year by 2025, increasing direct losses of coastal habitat from about 4000 km^2 (in the 1990s) to 8000 km^2 by 2025 (figures quoted by Tucker and Evans, 1997). Another significant threat to Mediterranean coastal wetlands comes from the combination of hunting and toxic pollution. Illegal hunting is commonplace in southern Europe, especially in autumn and winter. Lead poisoning in waterfowl occurs as the birds ingest spent gunshot, particularly in sediments which contain little grit. In the Camargue shot densities of up to 2 million/ha have been recorded and more than 50% of the gizzards of some duck species were found to contain lead shot (Pain, 1992). The shot may remain in the gizzard, or be ground down, dissolved by stomach acids and absorbed into the bloodstream. Although lead shot has been replaced by non-toxic alternatives in northern Europe there is no legislation for this in southern Europe (Tucker and Evans, 1997).

9.3.5 Conservation strategies for individual species

The existence of umbrella species whose conservation would also protect others which share the same habitat has already been referred to. Usually umbrella species are large mammals and birds, predators at the top of the food chain. By virtue of this position they are increasingly rare as their geographical resource bases are large. Many in the Mediterranean are already extinct, such as the lion. A few remain and have, in places, acquired an iconic status with special measures derived for their conservation. The example given here is the Iberian lynx (*Lynx pardinus*), threatened by habitat destruction and fragmentation. In addition, plant species conservation may be achieved through localised initiatives in Mediterranean botanic gardens, the equivalent perhaps of the captive breeding programme described for the lynx. Some consideration to the role of botanic gardens is also given in this section.

9.3.5a The Iberian lynx

The lynx has an IUCN endangered status. It is the most threatened European carnivore, confined to central and southwestern Spain and southern Portugal (Rodriguez and Delibes, 1992). The current population is about 1200 but with only 350 breeding females. This is divided between a number of locations. Some areas only contain about a dozen individuals, but the relatively large areas of the Sierra Morena and Montes de Toledo in Spain contain about 70–80% of the total population. Their diet is almost exclusively rabbit and although numerous today, the decline of rabbits due to myxomatosis in the past greatly reduced available prey. Other reasons for the decline in lynx include clearance of maquis since the 1940s. Lynx are nocturnal and rest during the daytime in thick heather scrub. Plantations of pines and eucalyptus have also reduced their habitat. Lynx have also been perceived as predators of livestock and shot for their pelts. The creation of reserves large enough to guarantee their survival is not feasible, therefore short-term actions have been proposed: prevention of further habitat loss by creating protected areas; encouraging their survival in the large central population; linking adjacent populations by natural corridors or artificial contact (lynx release, insemination); through reintroduction in southern Spain; and by captive breeding. Lynx are found in the Coto Doñana National Park and a captive breeding programme has been established; unfortunately many lynx are run over by cars within the park. There are also reports of tuberculosis in lynx in the park which could have spread from grazing cattle and might have a devastating effect on lynx numbers (Anon., 2000). The long-term survival of the lynx is unlikely, but the Spanish government is developing a national conservation strategy for it.

9.3.5b Botanic gardens

Botanic gardens have a role to play in the conservation of Mediterranean plant species. In the Mediterranean countries there are over 60 such gardens (Avishai, 1985). Not all of them originated with conservation as a main objective. Some are located in densely populated, and often polluted, urban areas and are frequently regarded as living museums of exotic and probably non-Mediterranean species. However, conservation of threatened Mediterranean plants is most feasible in Mediterranean climates and therefore in gardens in the region, rather than in botanic gardens of northern Europe and North America. To this end three basic actions are needed for the conservation of plants – obtaining the plants from the wild, establishing and maintaining populations in gardens and distributing propagated material as widely as possible between gardens. A part of the problem in achieving this lies in establishing the variety of species found in the region, in other words ensuring the use of the same nomenclature and taxonomy in all countries. Although *Flora Europaea* (for example Tutin *et al.*, 1992) is widely used there is still confusion and the need for a renewed consensus about nomenclature in the Mediterranean. As a result the MedChecklist of plants has been produced (for example Greuter *et al.*, 1989). In addition an electronic international database and information system for flowering plants, conifers and ferns is being developed for Europe, the Mediterranean region, Macronesia (the Azores and Canaries) and the Caucasus. This is known as the Euro+Med PlantBase. The intention is to provide a new consensus taxonomy for use by environmentalists, ecologists, planners, environmental legislators, biologists, horticulturalists and others. For the Mediterranean, the project is sponsored by OPTIMA (the Organization for the Phyto-taxonomic Investigation of the Mediterranean Area). A revised and detailed taxonomy of Mediterranean plants will form the basis of better conservation of species between and among botanical gardens.

9.4 Sustainable resource use

Through an analysis of the ecogeography of the Mediterranean region it is apparent that there is considerable heterogeneity – in terms of climate, geology, tectonic activity, land use, vegetation and animal communities. Nevertheless there are many unifying characteristics stemming from the same factors which give heterogeneity. As in other major biomes of the world, environmental issues are of increasing concern and the search for solutions is in many cases now an international co-operative effort. Most of the research cited in this book indicates that there is 'no fixed and unique recipe for addressing the Mediterranean type . . . problems. Neither is there a unique "technological" fix. Rather it is a matter of having a thorough understanding of the current state of the diverse environments' (Thornes, 1998b, p. 163). It is to be hoped that this book, drawing on research from across the Basin, and to some extent from other regions with mediterranean-type ecosystems, and from a variety of disciplines, has added a measure of this understanding. Sustainability is argued by many as the way forward. It is needed in order to achieve a better balance between the environment and economic activity already achieved in the wealthier nations of the Mediterranean. It is also relevant in the poorer countries where there is a very great need to raise living standards, especially in rural areas where population-induced pressures on the environment are considerable. One such conservation project, which has sustainability as a central theme, is the indigenous bulb propagation project, a co-operative effort between Turkish conservation groups and Flora and Fauna International. This is the subject of the last section.

9.4.1 Flora and Fauna International's indigenous propagation project

In May 1996 a special harvest festival took place in the village of Dumlugöze in Turkey. The celebration was for the first harvest of snowdrop and aconite bulbs grown in the fields around the village in the southern Taurus mountains (Byfield, 1996). For centuries people in the region had collected bulbs from the wild and exported these to bulb wholesalers in northern Europe, particularly the Dutch. In the late 1980s some 60 million bulbs were being exported each year and over half this trade was in snowdrops (*Galanthus elwesii* and *G. ikariae*). Other species of wild harvested bulbs included *Eranthis hyemalis*, *Anemone blanda*, *Leucojum aestivum* (snowflake), *Cyclamen cilicium*, *C. coum*, *C. hederifolium*, *Geranium tuberosum* and *Sternbergia lutea*. The bulbs were being sold in small packs, relatively cheaply, by wholesalers and then through supermarkets and garden centres in western Europe. The export was lucrative to the Turkish economy yet the harvest was unsustainable. All the species are listed under Appendix II of CITES (see Box 9.1).

In an attempt to reduce the trade Flora and Fauna International and the Turkish Society for the Protection of Nature, Dogal Hayati Koruma Dernegi, set up the Indigenous Plant Propagation Project in the early 1990s. Its aims are threefold: to reduce demand for wild collected bulbs by raising awareness of the decline in wild bulbs, both within the trade and with the final customers in western Europe; to encourage correct labelling of bulbs to distinguish between wild collected and cultivated or propagated bulbs; and to develop village-level cultivation of bulbs rather than reliance on collection from the wild. In a large measure the project has been a success. For the villages involved it has raised their profile, brought into use underused land around the villages, brought in external investment (although the low-level techniques used to produce a crop do not need a large financial investment), helped to increase the villagers' awareness of sustainable development and raised the self-esteem and confidence of the villagers, including women, who have more say in how bulbs should be cultivated and harvested. Some of the weaknesses of the project centre around the initial investment needed to create it in the first place, though this will decline as the project extends to other villages. There are also concerns that the land needed for bulb cultivation may compete with that needed for food production. As yet the level of income gained from cultivating bulbs is also low relative to the time required until the crop is ready for harvesting (three years from planting to harvesting) and the fact that collection of wild bulbs requires fewer resources than cultivating them. The main financial gain is not made by the village producer, who is paid approximately £1.50 per kilo for *Galanthus* while the Turkish exporters receive about £20 per kilo from the Dutch wholesalers. The cultivation of bulbs has not yet reached a sustainable volume.

Collection from the wild for export to western Europe and therefore illegal trade in bulbs is still taking place, but some trade in wild collected bulbs is allowed (even though many species are CITES-listed) under quotas set by Turkey. In 1998 the very rare *Cyclamen mirabile* was still being sold in UK supermarkets – it could only have come from the wild because it is not part of the propagation project. Despite publicity there remains a market for wild *Galanthus* bulbs for in many instances they are cheaper than propagated ones. As a consequence trade has expanded into the former states of the USSR, such as Georgia, Azerbaijan and Armenia. Fears obviously exist for the unregulated plunder of their biodiversity.

This chapter began with the export of bulbs from Turkey and has ended with the same subject, but as the last section has shown the nature of this trade changed dramatically in the 1990s. The Indigenous Propagation Project demonstrates that a traditional trade can be made into a relatively stable source of income based on the principles of sustainable resource use. The example could be applied to other plants, such as those with medicinal properties, and to other villages. Increased public awareness, in both Turkey and western Europe, should also provide an opportunity to show that co-operative projects between nations can be a benefit to the environment and should be encouraged in the future.

Appendix 1

A note on the taxonomy of plants and animals

The study of plants and animals anywhere in the world requires biologists, ecologists, geographers etc. to name the organisms that they come across. The system that has been used since the mid-eighteenth century is based on binomial nomenclature. This was first applied systematically by Carolus Linnaeus (Carl von Linné, 1707–1778), a Swedish naturalist who published two major works. The *Systema Naturae* (1735) was a classified list of plants, animals and minerals, and its tenth edition (1758) is the starting point for zoological nomenclature. *Species Plantarum* (1753) provides the basis for botanical nomenclature.

When a new organism is discovered its description is published and it is given a Latin name, written in italics. This name is in two parts – one part is the specific name peculiar to the species, the other part is the generic name designating the genus to which the species belongs. The generic name is given first and starts with an uppercase letter, while the specific name comes second, starting with a lowercase letter. For example the cork oak is *Quercus suber*. The name (or an abbreviation of it) of the person responsible for discovering and describing the species should strictly follow, for example *Quercus suber* L. (for Linnaeus himself) but this name is often omitted. Where subspecies or varieties have been identified, a trinomial nomenclature is used, with the third name indicating the subspecies or variety. Rules for the use of the binomial nomenclature are established by the International Code of Botanical Nomenclature for plants, and the International Code of Zoological Nomenclature for animals.

A main advantage of the binomial nomenclature is its universal acceptance among scientists. The use of Latin avoids any confusion that might occur with the use of local names. Understandably, the non-specialist often finds the Latin names impenetrable and hard to understand, and for many people (including, I suspect, many readers of this book) common (English) names would be preferable. However major confusions arise as a species may have more than one common name. For example, Rackham and Moody (1996) refer to *Quercus coccifera* as the prickly oak, while Huxley and Taylor (1977) and many others call this species the kermes oak. In addition to multiple common names, many species have no common names or, at least for the Mediterranean region, no English names. It is for this reason that binomial, Latin names are used in this book. However, wherever possible common (English) names are given when appropriate.

Table A1.1 Alternative names of selected plant families commonly found around the Mediterranean – a comparison of *Flora Europaea* (Tutin *et al.*, 1964–80) and Stace (1991)

Tutin *et al.*	Stace	Common (English) name
Compositae	Asteraceae	Daisy family
Cruciferae	Brassicaceae	Cabbage family
Gramineae	Poaceae	Grass family
Guttiferae (Hypericaceae)	Clusiaceae	St John's wort family
Labiatae	Lamiaceae	Mint family
Leguminosae	Fabaceae	Pea family
Umbelliferae	Apiaceae	Carrot family

Just as species are grouped into genera, so genera are grouped into families, and these into orders and then divisions of the plant or animal kingdoms. Family names are not written in italics. Taxonomy of plants and animals undergoes the normal process of continuing revision according to the relevant international codes. As a result names occasionally change. The effect of this is further confusion for the non-specialist when consulting reference books from different dates. For plants there has been a recent revision of names at the family level for the British Isles (Stace, 1991) based in part on the notion that family names should be related to the name of a genus within that family. In consequence some common family names have been changed. Those which are relevant to the Mediterranean are noted in Table A1.1; the names used by Stace are those referred to in this book. Occasionally this makes a turn of phrase quite awkward. For example, many leguminous plants are nitrogen-fixers. These are members of the pea family, formerly known as Leguminosae, but now known as Fabaceae. Similarly, members of the mint family (which are very common in the Mediterranean region) are often referred to as labiates. Formerly designated as Labiatae, the mint family is now known as Lamiaceae.

Appendix 2

Geological chronology

Table A2.1 Geological chronology (Harland, 1990)

Era	Period		Epoch	Age (millions of years)	Mediterranean events
Cenozoic	Quaternary		Holocene	0.01	
			Pleistocene	1.64	
					3.2 million years ago establishment of a mediterranean-type climate
	Tertiary	Neogene	Pliocene	5.2	Messinian salinity crisis 6.5–5 million years ago
			Miocene	23.3	Alpine orogeny during the Tertiary ended during the Miocene
		Palaeogene	Oligocene	35.4	
			Eocene	56.5	
			Palaeocene	65	
Mesozoic	Cretaceous			145.6	
	Jurassic			208	
	Triassic			245	
Palaeozoic	Permian			290	
	Carboniferous			362.5	Hercynian orogeny c. 300 million years ago (Variscan orogeny)
	Devonian			408.5	
	Silurian			439	Caledonian orogeny 400–500 million years ago
	Ordovician			510	
	Cambrian			570	
	Pre-Cambrian				

Ages at start of period.

232

Appendix 3

Calibration of the radiocarbon time scale

Radiocarbon dating of organic matter is based on the decay rate of carbon-14 (^{14}C) in the sample. The rate of radioactive decay of an isotope is known as the half-life, or the time taken for its radioactivity to be reduced by half. The decay starts (and the clock is to set to zero) at the moment when the organism dies. The time since death can then be determined by how much decay has occurred. ^{14}C decays to ^{14}N with a half-life of 5730 ± 40 years.

Radiocarbon dates are usually expressed as an age in years before present, or yr BP, often with an error function representing one standard deviation and an indication of the laboratory code for that determination. For example, a date from a sedimentary core taken in the Algarve, Portugal is Wk 7265 1250 ± 170 yr BP (Allen, 1999). This indicates that it was dated by the Radiocarbon Dating Laboratory, University of Waikato, New Zealand (sample number 7265) and that there is a 68% probability that the ^{14}C age of the sample lies between 1080 and 1420 yr BP. By convention BP is taken to be before AD 1950. This provides a fixed datum against which dates can be determined; without it the date of a sample might appear older merely because its age was determined more recently than a different sample of the same age but determined earlier. By using this convention an age expressed in yr BP can be turned into a BC/AD date by adding 1950 or, as a quick approximation, 2000. However, this gives only an approximate age.

Atmospheric levels of ^{14}C have not been constant during the Holocene. This has become apparent through the measurements of ^{14}C in individual tree rings of known calendrical age – their dates do not agree. For samples older than about 2500 years there is a systematic divergence between ^{14}C and tree-ring ages; ^{14}C dates significantly underestimate actual ages for material older than 2500 years. The discrepancy for material 5000 years old amounts to some 600 years (12%). This means than radiocarbon ages need to be calibrated to give true, or calendar, years. Calibration programs exist for the Holocene and allow radiocarbon ages to be transformed into calendar ages, expressed as Cal. yr BP (for calendar years before present – still 1950). Table A3.1 provides a summary conversion for changing uncalibrated dates (expressed as yr BP) into calibrated dates (expressed as Cal. yr BP) and vice versa. These are, however, only crude approximations; changes in atmospheric concentrations of ^{14}C are neither smooth nor consistent over time. This means that a radiocarbon date might actually have more than one calendar age.

Further details of this and other problems associated with radiocarbon dating can be found in Roberts (1998).

Table A3.1 Calibration table for radiocarbon ages (Roberts, 1998)

^{14}C age (yr BP)	Calendar age (Cal. yr BP) intercepts and range	Calendar age (Cal. yr BP)	^{14}C age equivalent (uncalibrated yr BP)
500	519	500	430
1000	928	1000	1120
1500	1354	1500	1600
2000	1938	2000	2060
2500	(2709)–2581–(2509)	2500	2450
3000	(3205)–3187–(3169)	3000	2900
3500	(3817)–3792–(3725)	3500	3320
4000	4436	4000	3670
4500	(5246)–5122–(5056)	4500	4010
5000	5731	5000	4430
5500	6294	5500	4790
6000	(6854)–6821–(6814)	6000	5260
6500	7384	6500	5720
7000	7787	7000	6130
7500	(8313)–8220–(8218)	7500	6700
8000	(8950)–8784–(8767)	8000	7260
8500	(9480)–9474–(9456)	8500	7760
9000	9979	9000	8110
9500	(10 537)–10 513–(10 482)	9500	8560
10 000	(11 700)–11 500–(11 000)	10 000	9040
11 000	12 917	11 000	9800
12 000	13 992	12 000	10 230
13 000	15 437	13 000	11 090
14 000	16 792	14 000	12 000
15 000	17 916	15 000	12 740
16 000	18 876	16 000	13 380
17 000	20 105	17 000	14 180
18 000	21 484	18 000	15 070
		19 000	16 120
		20 000	16 920
		21 000	17 630
		22 000	18 410

References

Abbott, J., Baker, M., Bisong, P. and 16 others (1992) Dehesa: its potential for conservation in Cabañeros Natural Park. *Discussion Papers in Conservation*, no. 56, Ecology and Conservation Unit, University College, London.

Adams, J., Maslin, M. and Thomas, E. (1999) Sudden climate transitions during the Quaternary. *Progress in Physical Geography*, **23**, 1–36.

Agee, J.K. (1998) Fire and pine ecosystems, in Richardson, D.M. (ed.), *Ecology and Biogeography of* Pinus, Cambridge University Press, Cambridge.

Ajtay, G.L., Ketner, P. and Duvigneaud, P. (1979) Terrestrial primary production and phytomass, in Bolin, B., Degens, E.T., Kempe, S. and Ketner, P. (eds), *The Global Carbon Cycle*, Wiley, New York.

Alcalá, L. (1999) Spanish steps towards geoconservation. *Earth Heritage*, no. 11, 14–15.

Allen, H.D. (1990) A postglacial record from the Kopais Basin, Greece, in Bottema, S., Entjes-Nieborg, G. and van Zeist, W. (eds), *Man's Role in the Shaping of the Eastern Mediterranean Landscape*, Balkema, Rotterdam.

Allen, H.D. (1996) Mediterranean environments, in Adams, W.M., Goudie, A.S. and Orme, A.R. (eds), *The Physical Geography of Africa*, Oxford University Press, Oxford.

Allen, H.D. (1997) The environmental conditions of the Kopais Basin, Boeotia during the Postglacial with special reference to the Mycenaean period, in Bintliff, J.L. (ed.), *Recent Developments in the History and Archaeology of Central Greece. Proceedings of the Sixth International Boeotian Conference*, British Archaeological Reports, **666**, 39–58, Archaeopress, Oxford.

Allen, H.D. (1999) Wetland ecosystem changes: Boca do Rio, Algarve, Portugal. *International Quaternary Association XV International Congress*, Abstracts, 3–11 August 1999, Durban, South Africa.

Andrade, C. (1992) Tsunami generated forms in the Algarve barrier islands (South Portugal). *Science of Tsunami Hazards*, **10**, 21–33.

Andrés, P., Mateos, E. and Ascaso, C. (1999) Soil arthropods, in Rodà, F., Retana, J., Gracia, C.A. and Bellot, J. (eds), *Ecology of Mediterranean Evergreen Oak Forests*, Springer-Verlag, Berlin.

Anon. (2000) Missing lynx. *New Scientist*, no. 2231, 15.

Archbold, O.W. (1995) *Ecology of World Vegetation*, Chapman & Hall, London.

Arnell, N.W. (1999) The effect of climate change on hydrological regimes in Europe: a continental perspective. *Global Environmental Change*, **9**, 5–23.

Asioli, A., Trincardi, F., Lowe, J.J. and Oldfield, F. (1999) Short-term climate changes during the Last Glacial–Holocene transition: comparison between Mediterranean records and the GRIP event stratigraphy. *Journal of Quaternary Science*, **14**, 373–381.

Aura, J.E., Villaverde, V., Morales, M.G., Sainz, C.G., Zilhao, J. and Straus, L.G. (1998) The Pleistocene–Holocene transition in the Iberian Peninsula: continuity and change in human adaptations. *Quaternary International*, **49/50**, 87–103.

Avishai, M. (1985) The role of Mediterranean botanic gardens in the maintenance of living conservation-oriented collections, in Gómez-Campo, C. (ed.), *Plant Conservation in the Mediterranean Area*, Junk, Dordrecht.

Bailey, G. (1997) The Klithi project: history, archaeology and structures of investigations, in Bailey, G (ed.), *Klithi: Palaeolithic Settlement and Quaternary Landscapes in Northwest Greece*, vol. 1, McDonald Institute for Archaeological Research, Cambridge.

Barbero, M., Bonin, G., Loisel, R. and Quézel, P. (1990) Changes and disturbances of forest ecosystems caused by human activities in the western part of the Mediterranean basin. *Vegetatio*, **87**, 151–173.

Barbero, M., Loisel, R., Quézel, P., Richardson, D.M. and Romane, F. (1998) Pines of the Mediterranean Basin, in Richardson, D.M. (ed.), *Ecology and Biogeography of Pinus*, Cambridge University Press, Cambridge.

Barker, G.W. and Hunt, C.O. (1995) Quaternary valley floor erosion and alluviation in the Bifferno Valley, Molise, Italy: the role of tectonics, climate, sea level change, and human activity, in Lewin, J., Macklin, M.G. and Woodward, J.C. (eds), *Mediterranean Quaternary River Environments*, Balkema, Rotterdam.

Barry, R.G. and Chorley, R.J. (1998) *Atmosphere, Weather and Climate*, 7th edition, Routledge, London.

Bech, J., Rustullet, J., Garrigó, J., Tobías, F.J. and Marínez, R. (1997) The iron content of some red Mediterranean soils from northeast Spain and its pedogenic significance. *Catena*, **28**, 211–229.

Behre, K.-E. (1990) Some reflections on anthropogenic indicators and the record of prehistoric occupation phases in pollen diagrams from the Near East, in Bottema, S., Entjes-Nieborg, G. and van Zeist, W. (eds), *Man's Role in the Shaping of the Eastern Mediterranean Landscape*, Balkema, Rotterdam.

Beniston, M. and Tol, R.S.J. (1998) Europe, in Watson, R.T., Zinyowera, M.C., Moss, R.H. and Dokken, D.J. (eds), *The Regional Impacts of Climate Change: An Assessment of Vulnerability*, (Intergovernmental Panel on Climate Change) Cambridge University Press, Cambridge.

Bennett, K.D., Tzedakis, P.C. and Willis, K.J. (1991) Quaternary refugia of north European trees. *Journal of Biogeography*, **18**, 103–115.

Bennett, R.J. (1997) Open systems – closed systems: a Portuguese vignette, in Stoddart, D.R. (ed.), *Process and Form in Geomorphology*, Routledge, London.

Biger, G. and Liphschitz, N. (1991) The recent distribution of *Pinus brutia*: a reassessment based on dendroarchaeological and dendrohistorical evidence from Israel. *The Holocene*, **1**, 157–161.

Bilandzija, J., Frankovic, M. and Kaucic, D. (1998) The Croatian Adriatic coast, in Conacher, A.J. and Sala, M. (eds), *Land Degradation in Mediterranean Environments of the World*, Wiley, Chichester.

Bintliff, J.L. (1977) *Natural Environment and Human Settlement in Prehistoric Greece*, British Archaeological Report Supplementary Series, no. 28, BAR, Oxford.

Birkeland, P.W. (1984) *Soils and Geomorphology*, Oxford University Press, Oxford.

Blondel, J. (1981) Structure and dynamics of bird communities in Mediterranean habitats, in di Castri, F., Goodall, D.W. and Specht, R. (eds), *Ecosystems of the World*, vol. 11 *Mediterranean-type Shrublands*, Elsevier, Amsterdam.

Blondel, J. (1991) Invasions and range modifications of birds in the Mediterranean Basin, in Groves, R.H. and di Castri, F. (eds), *Biogeography of Mediterranean Invasions*, Cambridge University Press, Cambridge.

Blondel, J., Vuilleumier, F., Marcus, L.F. and Terouanne, E. (1984) Is there ecomorphological convergence among mediterranean bird communities of Chile, California and France? *Evolutionary Biology*, **18**, 141–213.

Blondel, J. and Aronson, J. (1995) Biodiversity and ecosystem function in the Mediterranean Basin: human and non-human determinants, in Davis, G.W. and Richardson, D.M. (eds), *Mediterranean-type Ecosystems: The Function of Biodiversity*, Springer-Verlag, Berlin.

Blumler, M.A. (1993) Successional pattern and landscape sensitivity in the Mediterranean Near East, in Thomas, D.S.G. and Allison, R.J. (eds), *Landscape Sensitivity*, Wiley, Chichester.

Blumler, M.A. (1996) Ecology, evolutionary theory and agricultural origins, in Harris, D. (ed.), *The Origins and Spread of Agriculture and Pastoralism in Eurasia*, UCL Press, London.

Boero, V. and Schwertmann, U. (1989) Iron oxide mineralogy of terra rossa and its genetic implications. *Geoderma*, **44**, 319–327.

Bond, G., Heinrich, H., Broecker, W. and 11 others (1992) Evidence for massive discharge of icebergs into the North Atlantic ocean during the last glaciation. *Nature*, **360**, 245–249.

Bond, W.J. (1995) Effects of global change on plant–animal synchrony: implications for pollination and seed dispersal in Mediterranean habitats, in Moreno, J.M. and Oechel, W.C. (eds), *Global Change and Mediterranean-type Ecosystems*, Springer-Verlag, New York.

Bottema, S. (1995) The Younger Dryas in the eastern Mediterranean. *Quaternary Science Reviews*, **14**, 883–891.

Bottema, S. and Woldring, H. (1990) Anthropogenic indicators in the pollen record of the Eastern Mediterranean, in Bottema, S., Entjes-Nieborg, G. and van Zeist, W. (eds), *Man's Role in the Shaping of the Eastern Mediterranean Landscape*, Balkema, Rotterdam.

Bottner, P., Coûteaux, M.M. and Vallejo, V.R. (1995) Soil organic matter in Meterranean-type ecosystems and global climatic changes: a case study – the soils of the Mediterranean Basin, in Moreno, J.M. and Oechel, W.C. (eds), *Global Change and Mediterranean-type Ecosystems*, Springer-Verlag, New York.

Brandt, C.J. and Thornes, J.B. (eds) (1996) *Mediterranean Desertification and Land Use*, Wiley, Chichester.

Braudel, F. (1972, 1973) *The Mediterranean and the Mediterranean World in the Age of Philip II*, vols 1 and 2, Collins, London.

Braun-Blanquet, J. (1932) *Plant Sociology*, McGraw-Hill, New York.

Bridges, E.M. (1978) *World Soils*, 2nd edition, Cambridge University Press, Cambridge.

Britton, R.H. and Johnson, A.R. (1987) An ecological account of a Mediterranean salina: the Salin de Giraud, Camargue (S. France). *Biological Conservation*, **42**, 185–230.

Broecker, W. (1994) Massive iceberg discharges as triggers for global climate change. *Nature*, **372**, 421–424.

Burnie, D. (1995) *Wild Flowers of the Mediterranean*, Dorling Kindersley, London.

Byfield, A. (1996) Reaping the rewards. *The Garden*, November, The Royal Horticultural Society, London.

Cain, S. (1950) Lifeforms and phytoclimate. *Botanical Review*, **16**, 1–32.

Cantù, V. (1977) The climate of Italy, in Wallén, C.C. (ed.), *World Survey of Climatology*, vol. 6 *Central and Southern Europe*, Elsevier, Amsterdam.

Chambers, F. (1999) 'Global warming': new perspectives from palaeoecology and solar science. *Geography*, **83**, 266–271.

Chapman, J., Shiel, R. and Batovic, S. (1996) *The Changing Face of Dalmatia*, Leicester University Press, Leicester.

Chester, D.K. and Duncan, A.M. (1982) The interaction of volcanic activity in Quaternary times upon the evolution of the Alcantara and Simento rivers, Mount Etna, Sicily. *Catena*, **9**, 319–342.

Chester, D.K. and James, P.A. (1991) Holocene alluviation in the Algarve, southern Portugal: the case for an anthropogenic cause. *Journal of Archaeological Science*, **18**, 73–87.

Chester, R., Baxter, G.G., Behairy, A.K.A., Connor, K., Cross, D., Elderfield, H. and Padgham, R.C. (1977) Soil-sized eolian dusts from the lower troposphere of the eastern Mediterranean Sea. *Marine Geology*, **24**, 201–217.

Cheylan, G. (1991) Patterns of Pleistocene turnover, current distribution and speciation among Mediterranean mammals, in Groves, R.H. and di Castri, F. (eds), *Biogeography of Mediterranean Invasions*, Cambridge University Press, Cambridge.

Christensen, N.L. (1994) The effects of fire on physical and chemical properties of soils in Mediterranean-climate shrublands, in Moreno, J.M. and Oechel, W.C. (eds), *The Role of Fire in Mediterranean-type Ecosystems*, Springer-Verlag, New York.

Chuvieco, E., Deshayes, M., Stach, N., Cocero, D. and Riaño, D (1999) Short-term fire risk: foliage moisture content estimation from satellite data, in Chiviuco, E. (ed.), *Remote Sensing of Large Wildfires in the European Mediterranean Basin*, Springer-Verlag, Berlin.

Clark, S.C., Puigdefábregas, J. and Woodward, I. (1998) Aspects of the ecology of the shrub-winter annual communities, in Mairota, P., Thornes, J.B. and Geeson, N. (eds), *Atlas of Mediterranean Environments in Europe: The Desertification Context*, Wiley, Chichester.

Clements, F.E. (1916) *Plant Succession: An Analysis of the Development of Vegetation Cover*, Carnegie Institute, Washington, DC.

Cody, M.L. and Mooney, H.A. (1978) Convergence versus non-convergence in Mediterranean-climate ecosystems. *Annual Review of Ecology and Systematics*, **9**, 265–321.

Collier, R.E.Ll., Leeder, M.R. and Jackson, J.A. (1995) Quaternary drainage development, sediment fluxes and extensional tectonics in Greece, in Lewin, J., Macklin, M.G. and Woodward, J.C. (eds), *Mediterranean Quaternary River Environments*, Balkema, Rotterdam.

Conacher, A.J. and Sala, M. (eds) (1998) *Land degradation in Mediterranean Environments of the World*, Wiley, Chichester.

Corte-Real, J., Sorani, R. and Conte, M. (1998) Climate change, in Mairota, P., Thornes, J.B. and Geeson, N. (eds), *Atlas of Mediterranean Environments in Europe: The Desertification Context*, Wiley, Chichester.

Cowling, R.M. (1992) (ed.) *Fynbos: Nutrients, Fire and Diversity*, Oxford University Press, Cape Town.

Cowling, R.M., Rundel, P.W., Lamont, B.B., Arroyo, M.K. and Arianoutsou, M. (1996) Plant diversity in mediterranean-climate regions. *Trends in Ecology and Evolution*, **11**, 362–366.

Cox, C.B. and Moore, P.D. (1993) *Biogeography: An Ecological and Evolutionary Approach*, 5th edition, Blackwell, Oxford.

Cruzado, A. (1985) Chemistry of Mediterranean waters, in Margalef, R. (ed.), *The Western Mediterranean*, Pergamon, Oxford.

Damianos, D., Dimara, E., Hassapoyannes, K. and Skuras, D. (1998) *Greek Agriculture in a Changing International Environment*, Ashgate, Aldershot.

Dardis, G. and Smith, B. (1997) Coastal zone management, in King, R., Proudfoot, L. and Smith, B. (eds), *The Mediterranean: Environment and Society*, Arnold, London.

David, J.-F., Devernay, S., Loucougaray, G. and Le Floc'h, E. (1999) Belowground biodiversity in a Mediterranean landscape: relationships between saprophagous macroarthropod communities and vegetation structure. *Biodiversity and Conservation*, **8**, 753–767.

Davidson, D.A. (1980) Erosion in Greece during the first and second millennia BC, in Cullingford, R.A., Davidson, D.A. and Lewin, J. (eds), *Timescales in Geomorphology*, Wiley, Chichester.

Davis, F.W., Borchert, M.I. and Odion, D.C. (1989) Establishment of microscale vegetation patterns in maritime chaparral after fire. *Vegetatio*, **84**, 53–67.

Davis, G.W. and Richardson, D.M. (eds) (1995) *Mediterranean-type Ecosystems: The Function of Biodiversity*, Springer-Verlag, Berlin.

Davis, G.W. and Rutherford, M.C. (1995) Ecosystem function of biodiversity: can we learn from the collective experience of MTE research? in Davis, G.W. and Richardson, D.M. (eds), *Mediterranean-type Ecosystems: The Function of Biodiversity*, Springer-Verlag, Berlin.

Dawson, A.G., Hindson, R., Andrade, C., Freitas, C., Parish, R. and Bateman, M. (1995) Tsunami sedimentation associated with the Lisbon earthquake of 1 November AD 1755: Boca do Rio, Algarve, Portugal. *The Holocene*, **5**, 209–215.

Dedkhov, A.P. and Moszherin, V.I. (1992) Erosion and sediment yield in mountain regions of the world, in *Erosion, Debris Flow and Environment in Mountain Regions*, Proceedings of the Chengdu Symposium, July 1992, IAHS Publication no. 209, IAHS, Wallingford.

Delano-Smith, C. (1979) *Western Mediterranean Europe*, Academic Press, London.

di Castri, F. (1973) Soil animals in latitudinal and topographical gradients of Mediterranean ecosystems, in di Castri, F. and Mooney, H.A. (eds), *Mediterranean Type Ecosystems: Origin and Structure*, Springer-Verlag, Berlin.

di Castri, F. (1981) Mediterranean shrublands of the world, in di Castri, F., Goodall, D.W. and Specht, R. (eds), *Ecosystems of the World*, vol. 11 *Mediterranean-type Shrublands*, Elsevier, Amsterdam.

di Castri, F. (1991) The biogeography of animal invasions, in Groves, R.H. and di Castri, F. (eds), *Biogeography of Mediterranean Invasions*, Cambridge University Press, Cambridge.

di Castri, F. and Mooney, H.A. (eds) (1973) *Mediterranean Type Ecosystems: Origin and Structure*, Springer-Verlag, Berlin.

di Castri, F., Goodall, D.W. and Specht, R. (eds) (1981) *Ecosystems of the World*, vol. 11 *Mediterranean-type Shrublands*, Elsevier, Amsterdam.

Domínguez Lozano, F., Galicia Herbada, D., Moreno Rivero, L., Moreno Saiz, J.C. and Sainz Ollero, H. (1996) Threatened plants in Peninsular and Balearic Spain: a report based on the EU Habitats Directive. *Biological Conservation*, **76**, 123–133.

Douglas, T.D., Kirkby, S.J., Critchley, R.W. and Park, G.J. (1994) Agricultural terrace abandonment in the Alpujarra, Andalucia, Spain. *Land Degradation and Rehabilitation*, **5**, 281–291.

Dunford, M. (1997) Mediterranean economies: the dynamics of uneven development, in King, R., Proudfoot, L. and Smith, B. (eds), *The Mediterranean: Environment and Society*, Arnold, London.

Durrell, L. (1978) *The Greek Islands*, Faber & Faber, London.

Edwards, P.J. and Abivardi, C. (1998) The value of biodiversity: where ecology and economy blend. *Biological Conservation*, **83**, 239–246.

Ellis, S. and Mellor, A. (1995) *Soils and Environment*, Routledge, London.

Emberger, L. (1930) La végétation de la région méditerranéenne. Essai d'une classification des groupements végétaux. *Revue Générale de Botanique*, **42**, 641–662; 705–721.

Embleton, C.E. (1984) *Geomorphology of Europe*, Macmillan, London.

Esau, K. (1965) *Plant Anatomy*, 2nd edition, Wiley, New York.

Fa, J.E. (1986) On the ecological status of the Barbary macaque *Macaca sylvanus* L. in North Morocco: habitat influences versus human impact. *Biological Conservation*, **35**, 215–258.

Fantechi, R. and Margaris, N.S. (1986) *Desertification in Europe*, Reidel, Dordrecht.

FAO/UNESCO (1978) *Soil Map of the World*, UNESCO, Paris.

Faulkner, H. and Hill, A. (1997) Forests, soils and the threat of desertification, in King, R., Proudfoot, L. and Smith, B. (eds), *The Mediterranean: Environment and Society*, Arnold, London.

Finlayson, M., Hollis, G.E. and Davis, D. (eds) (1992) *Managing Mediterranean Wetlands and their Birds*, International Waterfowl and Wetlands Research Bureau Special Publication No. 20, International Waterfowl and Wetlands Research Bureau, Gloucester.

Flemming, N.C. (1992) Predictions of relative coastal sea-level change in the Mediterranean based on archaeological, historical and tide-gauge data, in Jeftic, L., Milliman, J.D. and Sestini, G. (eds), *Climatic Change and the Mediterranean* (United Nations Environment Programme), Arnold, London.

Fox, M.D. (1990) Mediterranean weeds: exchanges of invasive plants between the five Mediterranean regions of the world, in di Castri, F., Hansen, A.J. and Debussche, M. (eds), *Biological Invasions in Europe and the Mediterranean Basin*, Kluwer, Dordrecht.

Frihy, O.E. and Khafagy, A.A. (1991) Climate and induced changes in relation to shoreline migration trends at the Nile delta promontories, Egypt. *Catena*, **18**, 197–211.

Frogley, M.R., Tzedakis, P.C. and Heaton, T.H.E. (1999) Climate variability in northwest Greece during the last interglacial. *Science*, **285**, 1886–1889.

Furlan, D. (1977) The climate of southeastern Europe, in Wallén, C.C. (ed.), *World Survey of Climatology*, vol. 6 *Climates of Central and Southern Europe*, Elsevier, Amsterdam.

Garcia Novo, F. and Merino, J. (1993) Dry coastal ecosystems of southwest Spain, in van der Maarel, E. (ed.), *Ecosystems of the World*, vol. 2A *Dry Coastal Ecosystems*, Elsevier, Amsterdam.

Garcia Perez, J.D., Charlton, C. and Ruiz, P.M. (1995) Landscape changes as visible indicators in the social, economic and political process of soil erosion: a case study of the municipality of Puebla de Valles (Guadalajara province), Spain. *Land Degradation and Rehabilitation*, **6**, 149–161.

Gardiner, M.J. and Ryan, P. (1962) Relic soils on limestone in Ireland. *Irish Journal of Agricultural Research*, **1**, 181–188.

Garnsey, P. (1983) Grain for Rome, in Garnsey, P., Hopkins, K. and Whittaker, C.R. (eds), *Trade in the Ancient Economy*, Chatto and Windus, London.

Giles, B.D. and Balfoutis, C.J. (1990) The Greek heatwaves of 1987 and 1988. *International Journal of Climatology*, **10**, 505–517.

Gilman, A. and Thornes, J.B. (1985) *Land Use and Prehistory in South-east Spain*, Allen & Unwin, London.

Golley, F.B. (1993) *A History of the Ecosystem Concept in Ecology*, Yale University Press, New Haven, CT.

Gómez-Campo, C. and Herranz-Sanz, J.M. (1993) Conservation of Iberian endemic plants: the botanical reserve of la Encantada (Villarrobledo, Albacete, Spain). *Biological Conservation*, **64**, 155–160.

Goodess, C.M., Mariani, L., Palutikof, J.P., Menichini, V. and Minardi, G.P. (1998) Estimating future climates in the Mediterranean, in Mairota, P., Thornes, J.B. and Geeson, N. (eds), *Atlas of Mediterranean Environments in Europe: The Desertification Context*, Wiley, Chichester.

Got, H., Aloisi, J.-C. and Monaco, A. (1985) Sedimentary processes in Mediterranean deltas and shelves, in Stanley, D. and Wezel, F.C. (eds), *Geological Evolution of the Mediterranean Basin* Springer-Verlag, Berlin.

Goudie, A.S. (1995) *The Changing Earth: Rates of Geomorphological Processes*, Oxford University Press, Oxford.

Goudie, A.S., Atkinson, B.W., Gregory, K.J., Simmons, I.G., Stoddart, D.R. and Sugden, D. (eds) (1994) *The Encyclopedic Dictionary of Physical Geography*, 2nd edition, Blackwell, Oxford.

Gouyon, P.H., Vernet, P., Guillerm, J.L. and Valdeyron, G. (1986) Polymorphisms and environment: the adaptive value of the oil polymorphisms in *Thymus vulgaris* L. *Heredity*, **57**, 59–66.

Graham, B. (1997) The Mediterranean in the medieval and renaissance world, in King, R., Proudfoot, L. and Smith, B. (eds), *The Mediterranean: Environment and Society*, Arnold, London.

Grenon, M. and Batisse, M. (1989) *Futures for the Mediterranean Basin: The Blue Plan*, Oxford University Press, Oxford.

Greuter, W. (1994) Extinctions in Mediterranean areas. *Philosophical Transactions of the Royal Society of London Series B*, **344**, 41–46.

Greuter, W., Burdet, H.M. and Long, G. (1989) *Med-Checklist: A Critical Inventory of the Vascular Plants of the Circum-Mediterranean Countries*, vol. 4, Conservatoire et Jardin botaniques de la Ville de Genève & Secretariat Med-Checklist, Geneva.

Griffiths, J.F. (1972) The Mediterranean zone, in Griffiths, J.F. (ed.), *World Survey of Climatology*, vol. 10 *Climates of Africa*, Elsevier, Amsterdam.

Grove, A.T. (1986) The scale factor in relation to the processes involved in 'desertification' in Europe, in Fantechi, R. and Margaris, N.S. (eds), *Desertification in Europe*, Reidel, Dordrecht.

Grove, A.T. (1997) Classics in physical geography revisited. *Progress in Physical Geography*, **21**, 251–256.

Grove, A.T. and Rackham, O. (1993) Threatened landscape in the Mediterranean: examples from Crete. *Landscape and Urban Planning*, **24**, 279–292.

Grove, A.T. and Rackham, O. (1998) History of Mediterranean land use, in Mairota, P., Thornes, J.B. and Geeson, N. (eds), *Atlas of Mediterranean Environments in Europe: The Desertification Context*, Wiley, Chichester.

Grove, A.T. and Rackham, O. (2000) *The Nature of Mediterranean Europe: An Historical Ecology*, Yale University Press, New Haven, CT.

Grove, J.M. (1988) *The Little Ice Age*, Methuen, London.

Groves, R.H. (1991) The biogeography of mediterranean plant invasions, in Groves, R.H. and di Castri, F. (eds), *Biogeography of Mediterranean Invasions*, Cambridge University Press, Cambridge.

Groves, R.H. and di Castri, F. (1991) (eds) *Biogeography of Mediterranean Invasions*, Cambridge University Press, Cambridge.

Gunn, J. (1986) Solute processes and karst landforms, in Trudgill, S.T. (ed.), *Solute Processes*, Wiley, Chichester.

Hardy, D.A. and Renfrew, A.C. (eds) (1990) *Thera and the Aegean World*, vol. 3 *Chronology*, Thera Foundation, London.

Harland, W.B. (1990) *A Geological Time Scale*, Cambridge University Press, Cambridge.

Harley, P.C. (1995) Modeling leaf level effects of elevated CO_2 on Mediterranean sclerophylls, in Moreno, J.M. and Oechel, W.C. (eds), *Global Change and Mediterranean-type Ecosystems*, Springer-Verlag, New York.

Harris, D.R. (1996) Introduction: themes and concepts in the study of early agriculture, in Harris, D. (ed.), *The Origins and Spread of Agriculture and Pastoralism in Eurasia*, UCL Press, London.

Harrison, R.J. (1996) Arboriculture in southwest Europe: *dehesas* as managed woodlands, in Harris, D. (ed.), *The Origins and Spread of Agriculture and Pastoralism in Eurasia*, UCL Press, London.

Harrison, S.P. and Digerfeldt, G. (1993) European lakes as palaeohydrological and palaeoclimatic indicators. *Quaternary Science Reviews*, **12**, 233–248.

Herrera, C.M. (1992) Historical effects and sorting processes as explanations for contemporary ecological patterns: character syndromes in Mediterranean woody plants. *The American Naturalist*, **140**, 421–446.

Heun, M., Schäfer-Pregl, R., Klawan, D. and 4 others (1997) Site of einkorn wheat domestication identified by DNA fingerprinting. *Science*, **278**, 1312–1314.

Heywood, V.H. and Watson, W.T. (eds) (1995) *Global Biodiversity Assessment* (United Nations Environment Programme), Cambridge University Press, Cambridge.

Hilbert, D.W. and Canadell, J. (1995) Biomass partitioning and resource allocation of plants from Mediterranean-type ecosystems: possible responses to elevated atmospheric CO_2, in Moreno, J.M. and Oechel, W.C. (eds), *Global Change and Mediterranean-type Ecosystems*, Springer-Verlag, New York.

Hindson, R.A. and Andrade, C. (1999) Sedimentation and hydrodynamic processes associated with the tsunami generated by the 1755 Lisbon earthquake. *Quaternary International*, **56**, 27–38.

HMSO (1962) *Weather in the Mediterranean*, HMSO, London.

Hobbs, R.J., Richardson, D.M. and Davis, G.W. (1995) Mediterranean-type ecosystems: opportunities and constraints for studying the function of biodiversity, in Davis, G.W and Richardson, D.M. (eds), *Mediterranean-type Ecosystems: The Function of Biodiversity*, Springer-Verlag, Berlin.

Holdgate, M. (1994) Protected areas in the future: the implications of change, and the need for new policies. *Biodiversity and Conservation*, **3**, 406–410.

Hollis, G.E. (1992) The causes of wetland loss and degradation in the Mediterranean, in Finlayson, M., Hollis, G.E. and Davis, D. (eds), *Managing Mediterranean Wetlands and their Birds*, International Waterfowl and Wetlands Research Bureau Special Publication No. 20, International Waterfowl and Wetlands Research Bureau, Gloucester.

Hopkins, K. (1983) Introduction, in Garnsey, P., Hopkins, K. and Whittaker, C.R. (eds), *Trade in the Ancient Economy*, Chatto and Windus, London.

Houghton, J. (1997) *Global Warming: The Complete Briefing*, 2nd edition, Cambridge University Press, Cambridge.

Houghton, J., Meiro Filho, L.G., Callander, B.A., Harris, N., Kattenberg, A. and Maskell, K. (1996) *Climate Change 1995: The Science of Climate Change*, IPCC, Cambridge University Press, Cambridge.

Hsü, K.J., Ryan, W.B.F. and Citta, M.B. (1973) Late Miocene desiccation of the Mediterranean Sea. *Nature*, **242**, 240–244.

Huggett, R.J. (1998) *Fundamentals of Biogeography*, Routledge, London.

Huntley, B. and Birks, H.J.B. (1983) *An Atlas of Past and Present Pollen Maps for Europe 0–13 000 Years Ago*, Cambridge University Press, Cambridge.

Huxley, A. and Taylor, W. (1977) *Flowers of Greece and the Aegean*, Chatto and Windus, London.

Inbar, M. (1998) The eastern Mediterranean, in Conacher, A.J. and Sala, M. (eds), *Land Degradation in Mediterranean Environments of the World*, Wiley, Chichester.

IUCN Species Survival Commission (1994) *IUCN Red List Categories*, IUCN Publications Services Unit, Cambridge.

James, P.A. and Chester, D.K. (1995) Soils of Quaternary river sediments in the Algarve, in Lewin, J., Macklin, M. and Woodward, J. (eds) *Mediterranean Quaternary River Environments*, Balkema, Rotterdam.

Jeftic, L., Milliman, J.D. and Sestini, G. (eds) (1992) *Climatic Change and the Mediterranean* (United Nations Environment Programme), Arnold, London.

Joffé, G. (1993) Introduction, in Joffé, G. (ed.), *North Africa: Nation, State, and Region*, Routledge, London.

Joffre, R. (1992) The dehesa: does this complex ecological system have a future?, in Teller, A., Mathy, P. and Jeffers, J.N.R. (eds), *Responses of Forest Ecosystems to Environmental Changes*, Elsevier, London.

Keeley, J.E. and Baer-Keeley, M. (1999) Role of charred wood, heat-shock, and light in germination of postfire phrygana species from the eastern Mediterranean Basin. *Israel Journal of Plant Sciences*, **47**, 11–16.

Kelletat, D. (1991) The 1550 BP tectonic event in the Eastern Mediterranean as a basis for assessing the intensity of shore processes. *Zeitschrift für Geomorphologie Suppl.-Bd*, **81**, 181–194.

King, G., Sturdy, D. and Bailey, G. (1997) The tectonic background to the Epirus landscape, in Bailey, G. (ed.), *Klithi: Palaeolithic Settlement and Quaternary Landscapes in Northwest Greece*, vol. 2, McDonald Institute for Archaeological Research, Cambridge.

King, R. (1997a) Introduction: an essay on Mediterraneanism, in King, R., Proudfoot, L. and Smith, B. (eds), *The Mediterranean: Environment and Society*, Arnold, London.

King, R. (1997b) Population growth: an avoidable crisis?, in King, R., Proudfoot, L. and Smith, B. (eds), *The Mediterranean: Environment and Society*, Arnold, London.

Kirkby, M.J., Bathurst, J.C., Woodward, I. and Thornes, J.B. (1998) Modelling the physical and biological systems, in Mairota, P., Thornes, J.B. and Geeson, N. (eds), *Atlas of Mediterranean Environments in Europe: The Desertification Context*, Wiley, Chichester.

Klaus, W. (1989) Mediterranean pines and their history. *Plant Systematics and Evolution*, **162**, 133–163.

Kosmos, C. and Danalatos, N. (1998) Land-use change and soil properties affecting desertification at the Sparta field site, Athens, Greece, in Mairota, P., Thornes, J.B. and Geeson, N. (eds), *Atlas of Mediterranean Environments in Europe: The Desertification Context*, Wiley, Chichester.

Kosmos, C., Danalatos, N., Cammeraat, L.H. and 20 others (1997) The effect of land use on runoff and soil erosion rates under Mediterranean conditions. *Catena*, **29**, 45–59.

Kraus, H. and Alkhalaf, A. (1995) Characteristic surface energy budgets for different climate types. *International Journal of Climatology*, **15**, 275–284.

Kruger, F.J., Mitchell, D.T. and Jarvis, J.U.M. (1983) *Mediterranean-type Ecosystems: the Role of Nutrients*, Springer-Verlag, Berlin.

Kummerow, J. (1973) Comparative anatomy of sclerophylls of mediterranean climatic areas, in di Castri, F. and Mooney, H.A. (eds), *Mediterranean Type Ecosystems: Origin and Structure*, Springer-Verlag, Berlin.

Kutiel, P. and Inbar, M. (1993) Fire impacts on soil nutrients and soil erosion in a Mediterranean pine forest plantation. *Catena*, **20**, 129–139.

Kutiel, P. and Naveh, Z. (1987) The effect of fire on nutrients in pine forest soil. *Plant and Soil*, **104**, 262–274.

Kutzbach, J.E. and Guetter, P.J. (1986) The influence of changing orbital parameters and surface boundary conditions on climate simulations of the past 18 000 years. *Journal of Atmospheric Sciences*, **43**, 1726–1759.

Kuzucuoglu, C. (1995) River response to Quaternary tectonics with examples from northwestern Anatolia, Turkey, in Lewin, J., Macklin, M.G. and Woodward, J.C. (eds), *Mediterranean Quaternary River Environments*, Balkema, Rotterdam.

Lamb, H.F., Eicher, U. and Switsur, V.R. (1989) An 18 000-year record of vegetation, lake-level and climatic change from Tigalmamine, Middle Atlas, Morocco. *Journal of Biogeography*, **16**, 65–74.

Lambeck, K. (1995) Late Pleistocene and Holocene sea-level change in Greece and southwestern Turkey: a separation of eustatic, isostatic and tectonic contributions. *Geophysical Journal International*, **122**, 1022–1044.

Laouina, A. (1998) North Africa, in Conacher, A.J. and Sala, M. (eds), *Land Degradation in Mediterranean Environments of the World*, Wiley, Chichester.

Le Floc'h, E. (1991) Invasive plants of the Mediterranean Basin, in Groves, R.H. and di Castri, F. (eds), *Biogeography of Mediterranean Invasions*, Cambridge University Press, Cambridge.

Le Houérou, H.N. (1981) Impact of man and his animals on Mediterranean vegetation, in di Castri, F., Goodall, D.W. and Specht, R. (eds), *Ecosystems of the World*, vol. 11 *Mediterranean-type Shrublands*, Elsevier, Amsterdam.

Le Houérou, H.N. (1990) Global change: population, land-use and vegetation in the Mediterranean Basin by the mid-21st century, in Paepe, R., Fairbridge, R.W. and Jelgersma, S. (eds), *Greenhouse Effect, Sea Level and Drought* (NATO Scientific Affairs Division), Kluwer, Dordrecht.

Le Maitre, D.C. (1998) Pines in cultivation: a global view, in Richardson, D.M. (ed.), *Ecology and Biogeography of Pinus*, Cambridge University Press, Cambridge.

Leigh Fermor, P. (1958) *Mani*, Penguin, Harmondsworth.

Leon, C., Lucas, G. and Synge, H. (1985) The value of information in saving threatened Mediterranean plants, in Gómez-Campo, C. (ed.), *Plant Conservation in the Mediterranean Area*, Junk, Dordrecht.

Lepart, J. and Debussche, M. (1991) Invasion processes as related to succession and disturbance, in Groves, R.H. and di Castri, F. (eds), *Biogeography of Mediterranean Invasions*, Cambridge University Press, Cambridge.

Lewthwaite, J.G. (1982) Acorns for ancestors: the prehistoric exploitation of woodlands in the west Mediterranean, in Bell, M. and Limbrey, S. (eds), *Archaeological Aspects of Woodland Ecology*, BAR International Series no. 146, BAR, Oxford.

Lindh, G. (1992) Hydrology and water resources impact of climate change, in Jeftic, L., Milliman, J.D. and Sestini, G. (eds), *Climatic Change and the Mediterranean* (United Nations Environment Programme), Arnold, London.

Linés Escardó, A. (1970) The climate of the Iberian Peninsula, in Wallén, C.C. (ed.), *World Survey of Climatology*, vol. 5 *Climates of Northern and Western Europe*, Elsevier, Amsterdam.

Llusia, J. and Penueles, J. (2000) Seasonal patterns of terpene content and emission from seven Mediterranean woody species in field conditions. *American Journal of Botany*, 87, 133–140.

López-Iborra, G. and Gil-Delgado, J.A. (1999) Composition and dynamics of the bird community, in Rodà, F., Retana, J., Gracia, C.A. and Bellot, J. (eds), *Ecology of Mediterranean Evergreen Oak Forests*, Springer-Verlag, Berlin.

Mabberley, D.J. and Placito, P.J. (1993) *Algarve Plants and Landscape*, Oxford University Press, Oxford.

Macklin, M.G., Lewin, J. and Woodward, J.C. (1995) Quaternary fluvial systems in the Mediterranean Basin, in Lewin, J., Macklin, M.G. and Woodward, J.C. (eds), *Mediterranean Quaternary River Environments*, Balkema, Rotterdam.

Macklin, M.G., Lewin, J. and Woodward, J.C. (1997) Quaternary river sedimentary sequences of the Voidomatis Basin, in Bailey, G. (ed.), *Klithi: Palaeolithic Settlement and Quaternary Landscapes in Northwest Greece*, vol. 2, McDonald Institute for Archaeological Research, Cambridge.

MacLeod, D.A. (1980) The origin of red Mediterranean soils in Epirus, Greece. *Journal of Soil Science*, 31, 125–136.

Mairota, P., Thornes, J.B. and Geeson, N. (eds) (1998) *Atlas of Mediterranean Environments in Europe: The Desertification Context*, Wiley, Chichester.

Mannion, A.M. (1995) *Agriculture and Environmental Change: Temporal and Spatial Dimensions*, Wiley, Chichester.

Mannion, A.M. (1997) *Global Environmental Change*, 2nd edition, Longman, Harlow.

Mannion, A.M. (1999) Domestication and origins of agriculture: an appraisal. *Progress in Physical Geography*, 23, 37–56.

Margaris, N.S. (1975) Effect of photoperiod on seasonal dimorphism of some mediterranean plants. *Berichte der Schweizerischen botanischen Gesellschaft*, 85, 96–102.

Margaris, N.S. (1977) Physiological and biochemical observations in seasonal dimorphic leaves of *Sarcopoterium spinosum* and *Phlomis fruticosa*. *Oecologia Plantarum*, 12, 343–350.

Margaris, N.S. (1981) Adaptive strategies in plants dominating mediterranean-type ecosystems, in di Castri, F., Goodall, D.W. and Specht, R. (eds), *Ecosystems of the World*, vol. 11 *Mediterranean-type Shrublands*, Elsevier, Amsterdam.

Marquez, R. and Alberch, P. (1995) Monitoring environmental change through amphibian populations, in Moreno, J.M. and Oechel, W.C. (eds), *Global Change and Mediterranean-type Ecosystems*, Springer-Verlag, New York.

Martin, P.S. (1984) Pleistocene overkill: the global model, in Martin, P.S. and Klein, P.G. (eds), *Quaternary Extinctions*, University of Arizona Press, Tucson, AZ.

Mather, A.E., Silva, P.G., Goy, J.L., Harvey, A.M. and Zazo, C. (1995) Tectonics versus climate: an example from late Quaternary aggradational and dissectional sequences of the Mula basin, southeast Spain, in Lewin, J., Macklin, M.G. and Woodward, J.C. (eds), *Mediterranean Quaternary River Environments*, Balkema, Rotterdam.

Maundrell, H. (1749) *A Journey from Aleppo to Jerusalem; at Easter 1679*, 7th edition, Meadows Booksellers, Oxford.

Mazzoleni, S., Lo Porto, A. and Blasi, C. (1992) Multivariate analysis of climatic patterns of the Mediterranean basin. *Vegetatio*, 98, 1–12.

McNeill, J.R. (1992) *The Mountains of the Mediterranean World*, Cambridge University Press, Cambridge.

Médail, F. and Quézel, P. (1997) Hot-spots analysis for conservation of biodiversity in the Mediterranean Basin. *Annals of the Missouri Botanical Garden*, **84**, 112–127.

Middleton, N. and Thomas, D.S.G. (1994) *Desertification: Exploding the Myth*, Wiley, Chichester.

Miglietta, F., Badiani, M., Bettarini, I., van Gardingen, P., Selvi, F. and Raschi, A. (1995) Preliminary studies of the long-term CO_2 response of Mediterranean vegetation around natural CO_2 vents, in Moreno, J.M. and Oechel, W.C. (eds), *Global Change and Mediterranean-type Ecosystems*, Springer-Verlag, New York.

Mooney, H.A. and Dunn, E.L. (1970) Convergent evolution of mediterranean-climate evergreen scleophyllous shrubs. *Evolution*, **24**, 292–303.

Moreno, J.M. and Oechel, W.C. (eds) (1994) *The Role of Fire in Mediterranean-type Ecosystems*, Springer-Verlag, New York.

Moreno, J.M. and Oechel, W.C. (eds) (1995) *Global Change and Mediterranean-type Ecosystems*, Springer-Verlag, New York.

Myers, N., Mittermeier, R.A., Mittermeier, C.G., da Fonseca, G.A.B. and Kent, J. (2000) Biodiversity hotspots for conservation priorities, *Nature*, **403**, 853–858.

Nahal, I. (1981) The Mediterranean climate from a biological viewpoint, in di Castri, F., Goodall, D.W. and Specht, R. (eds), *Ecosystems of the World*, vol. 11 *Mediterranean-type Shrublands*, Elsevier, Amsterdam.

Naveh, Z. (1974) Effects of fire in the Mediterranean region, in Kozlowski, T.T. and Ahlgren, C.E. (eds), *Fire and Ecosystems*, Academic Press, New York.

Naveh, Z. (1975) The evolutionary significance of fire in the Mediterranean region. *Vegetatio*, **9**, 199–206.

Naveh, Z. (1993) Red books for threatened Mediterranean landscapes: principles and first results of the Crete case study. *Landscape and Urban Planning*, **24**, 241–247.

Naveh, Z. (1994) The role of fire and its management in the conservation of Mediterranean ecosystems and landscapes, in Moreno, J.M. and Oechel, W.C. (eds), *The Role of Fire in Mediterranean-type Ecosystems*, Springer-Verlag, New York.

Naveh, Z. (1995) Conservation, restoration and research priorities for Mediterranean uplands threatened by global climate change, in Moreno, J.M. and Oechel, W.C. (eds), *Global Change and Mediterranean-type Ecosystems*, Springer-Verlag, New York.

Naveh, Z. and Lieberman, A.S. (1994) *Landscape Ecology: Theory and Applications*, Springer-Verlag, New York.

Nihlén, T. and Mattson, J.O. (1989) Studies in eolian dust. *Geografiska Annaler*, **71A**, 269–274.

Noss, R.F. (1990) Indicators for monitoring biodiversity: a hierarchical approach. *Conservation Biology*, 4, 355–364.

Oechel, W.C., Hastings, S.J., Vourlitis, G.L., Jenkins, M.A. and Hinkson, C.L. (1995) Direct effects of elevated CO_2 in chaparral and Mediterranean-type ecosystems, in Moreno, J.M. and Oechel, W.C. (eds), *Global Change and Mediterranean-type Ecosystems*, Springer-Verlag, New York.

Ojeda, F., Arroyo, J. and Marañon, T. (1995) Biodiversity components and conservation of Mediterranean heathlands in southern Spain. *Biological Conservation*, **72**, 61–72.

Pahl-Wostl, C. (1995) *The Dynamic Nature of Ecosystems: Chaos and Order Entwined*, Wiley, Chichester.

Pain, D. (1992) Lead poisoning in birds: a southern perspective, in Finlayson, M., Hollis, G.E. and Davis, D. (eds), *Managing Mediterranean Wetlands and their Birds*,

International Waterfowl and Wetlands Research Bureau Special Publication No. 20, International Waterfowl and Wetlands Research Bureau, Gloucester.

Papanastasis, V.P. and Kazaklis, A. (1998) Land use changes and conflicts in the Mediterranean-type ecosystems of western Crete, in Rundel, P.W., Montenegro, G. and Jaksic, F.M. (eds), *Landscape Disturbance and Biodiversity in Mediterranean-type Ecosystems*, Springer-Verlag, Berlin.

Parsons, J.J. (1962) The acorn-hog economy of the oak woodlands of southwestern Spain. *Geographical Review*, **52**, 211–235.

Pausas, J.G. and Vallejo, V.R. (1999) The role of fire in European Mediterranean ecosystems, in Chiviuco, E. (ed.), *Remote Sensing of Large Wildfires in the European Mediterranean Basin*, Springer-Verlag, Berlin.

Pavord, A. (1999) *The Tulip*, Bloomsbury, London.

Perelman, R. (1981) Perception of Mediterranean landscapes, particularly of maquis landscape, in di Castri, F., Goodall, D.W. and Specht, R. (eds), *Ecosystems of the World*, vol. 11 *Mediterranean-type Shrublands*, Elsevier, Amsterdam.

Perry, A. (1981) Mediterranean climate – a synoptic reappraisal. *Progress in Physical Geography*, **5**, 107–113.

Perry, A. (1997) Mediterranean climate, in King, R., Proudfoot, L. and Smith, B. (eds), *The Mediterranean: Environment and Society*, Arnold, London.

Piggot, C.D. (1962) Soil formation and development on carboniferous limestone of Derbyshire. *Journal of Ecology*, **50**, 145–156.

Piñol, J., Terradas, J. and Lloret, F. (1998) Climate warming, wildfire hazard and wildfire occurrence in coastal eastern Spain. *Climatic Change*, **38**, 345–357.

Pirazzoli, P.A. (1998) *Sea Level Changes: The Last 20 000 Years*, Wiley, Chichester.

Plato (1929) *Critias*, translated by Bury, R.G., Heinemann, London.

Poesen, J. and Bunte, K. (1996) Effects of rock fragments on desertification processes in Mediterranean environments, in Brandt, C.J. and Thornes, J.B. (eds), *Mediterranean Desertification and Land Use*, Wiley, Chichester.

Poesen, J., van Wesemael, B. and Bunte, K. (1998) Soils containing rock fragments and their response to desertification, in Mairota, P., Thornes, J.B. and Geeson, N. (eds), *Atlas of Mediterranean Environments in Europe: The Desertification Context*, Wiley, Chichester.

Poland, H.C., Hall, G.B. and Smith, M. (1996) Turtles and tourists: a hands-on experience of conservation for sixth formers from King's College, Taunton, on the Ionian Island of Zakynthos. *Journal of Biological Education*, **30**, 120–128.

Polunin, O. and Huxley, A. (1987) *Flowers of the Mediterranean*, Chatto and Windus, London.

Polunin, O. and Smythies, B.E. (1973) *Flowers of South-West Europe*, Oxford University Press, Oxford.

Pons, A. and Reille, M. (1988) The Holocene- and Upper Pleistocene pollen records from Padul (Granada, Spain): a new study. *Palaeogeography, Palaeoclimatology, Palaeoecology*, **66**, 243–263.

Pope, K. and van Andel., T.H. (1984) Late Quaternary alluviation and soil formation in the southern Argolid: its history, causes and archaeological implications. *Journal of Archaeological Science*, **11**, 281–306.

Pratt, J. and Funnell, D. (1997) The modernisation of Mediterranean agriculture, in King, R., Proudfoot, L. and Smith, B. (eds), *The Mediterranean: Environment and Society*, Arnold, London.

Proudfoot, L. (1997a) The Graeco-Roman Mediterranean, in King, R., Proudfoot, L. and Smith, B. (eds), *The Mediterranean: Environment and Society*, Arnold, London.

Proudfoot, L. (1997b) The Ottoman Mediterranean and its transformation, in King, R., Proudfoot, L. and Smith, B. (eds), *The Mediterranean: Environment and Society*, Arnold, London.

Puigdefábregas, J., Cueto, M., Domingo, F., Gutiérrez, L., Sánchez, G. and Solé, A. (1998) Rambla Honda, Tabernas, Almería, Spain, in Mairota, P., Thornes, J.B. and Geeson, N. (eds), *Atlas of Mediterranean Environments in Europe: The Desertification Context*, Wiley, Chichester.

Pullan, S. (1988) A survey of the past and present wetlands of the western Algarve, Portugal. *Liverpool Papers in Geography*, no. 2, Dept. of Geography, University of Liverpool.

Pye, K. (1992) Aeolian dust transport and deposition over Crete and adjacent parts of the Mediterranean Sea. *Earth Surface Process and Landforms*, **17**, 271–288.

Quézel, P. (1981) Floristic composition and phytosociological structure of sclerophyllous matorral around the Medtirranean, in di Castri, F., Goodall, D.W. and Specht, R. (eds), *Ecosystems of the World*, vol. 11 *Mediterranean-type Shrublands*, Elsevier, Amsterdam.

Quézel, P. (1985) Definition of the Mediterranean region and the origin of its flora, in Gómez-Campo, C. (ed.), *Plant Conservation in the Mediterranean Area*, Junk, Dordrecht.

Quézel, P. and Barbero, M. (1982) Definition and characterization of Mediterranean-type ecosystems. *Ecologia Méditerranea*, **8**, 15–29.

Rackham, O. (1982) Land-use and the native vegetation of Greece, in Bell, M. and Limbrey, S. (eds), *Archaeological Aspects of Woodland Ecology*, BAR International Series no. 146, BAR, Oxford.

Rackham, O. (1998) Savanna in Europe, in Kirby, K.J. and Watkins, C. (eds), *The Ecological History of European Forests*, CAB International, Wallingford.

Rackham, O. and Moody, J. (1996) *The Making of the Cretan Landscape*, Manchester University Press, Manchester.

Raven, P.H. (1973) The evolution of Mediterranean floras, in di Castri, F. and Mooney, H.A. (eds), *Mediterranean Type Ecosystems: Origin and Structure*, Springer-Verlag, Berlin.

Rendell, H. (1997) Earth surface processes in the Mediterranean, in King, R., Proudfoot, L. and Smith, B. (eds), *The Mediterranean: Environment and Society*, Arnold, London.

Retana, J., Espelta, J.M., Gracia, M. and Riba, M. (1999) Seedling recruitment, in Rodà, F., Retana, J., Gracia, C.A. and Bellot, J. (eds), *Ecology of Mediterranean Evergreen Oak Forests*, Springer-Verlag, Berlin.

Reyment, R.A. (1983) Palaeontological aspects of island biogeography: colonization and evolution of mammals on Mediterranean islands. *Oikos*, **41**, 299–306.

Roberts, N. (1998) *The Holocene*, 2nd edition, Blackwell, Oxford.

Robinson, J. (1991) Fire from space: global fire evaluation using infrared remote sensing. *International Journal of Remote Sensing*, **12**, 3–24.

Rodà, F., Retana, J., Gracia, C.A. and Bellot, J. (1999) Preface, in Rodà, F., Retana, J., Gracia, C.A. and Bellot, J. (eds), *Ecology of Mediterranean Evergreen Oak Forests*, Springer-Verlag, Berlin.

Rodriguez, A. and Delibes, M. (1992) Current range and status of the Iberian lynx *Felis pardina* Temminck, 1824 in Spain. *Biological Conservation*, **61**, 189–196.

Rognon, P. (1987) Late Quaternary climatic reconstruction for the Maghreb (North Africa). *Palaeogeography, Palaeoclimatology, Palaeoecology*, **58**, 11–34.

Rosenzweig, M.L. (1995) *Species Diversity in Space and Time*, Cambridge University Press, Cambridge.

Ross, J.D. and Sombrero, C. (1986) Environmental control of essential oil production in Mediterranean plants, in Harborne, J.B. and Tomas-Barberan, F.A. (eds), *Ecological Chemistry and Biochemistry of Plant Terpenoids*, Clarendon Press, Oxford.

Rossignol-Strick, M., Nesteroff, W., Olive, P. and Vergnaud-Grazzini, C. (1982) After the deluge: Mediterranean stagnation and sapropel formation. *Nature*, **295**, 105–110.

Roy, J. and Soulié, L. (1992) Germination and population dynamics of *Cistus* species in relation to fire. *Journal of Applied Ecology*, **29**, 647–655.

Ruffell, A. (1997) Geological evolution of the Mediterranean Basin, in King, R., Proudfoot, L. and Smith, B. (eds), *The Mediterranean: Environment and Society*, Arnold, London.

Ruíz, M. and Ruíz, J.P. (1986) Ecological history of transhumance in Spain. *Biological Conservation*, **37**, 73–86.

Rundel, P.W. (1998) Landscape disturbance in Mediterranean-type ecosystems: an overview, in Rundel, P.W., Montenegro, G. and Jaksic, F.M. (eds), *Landscape Disturbance and Biodiversity in Mediterranean-type Ecosystems*, Springer-Verlag, Berlin.

Rundel, P.W., Montenegro, G. and Jaksic, F.M. (eds) (1998) *Landscape Disturbance and Biodiversity in Mediterranean-type Ecosystems*, Springer-Verlag, Berlin.

Runnels, C.N. and Hansen, J. (1986) The olive in the prehistoric Aegean: the evidence for domestication in the Early Bronze Age. *Oxford Journal of Archaeology*, **5**, 299–308.

Ryan, W. and Pitman, W. (1998) *Noah's Flood*, Simon & Schuster, New York.

Sancho Comins, J., Bosque Sendra, J. and Moreno Sanz, F. (1993) Crisis and permanence of the traditional landscape in the central region of Spain. *Landscape and Urban Planning*, **23**, 155–166.

Schönfelder, I. and Schönfelder, P. (1990) *Wild Flowers of the Mediterranean,* Collins, London (originally published in German in 1984).

Schüle, W. (1993) Mammals, vegetation and the initial human settlement of the Mediterranean islands: a palaeoecological approach. *Journal of Biogeography*, **20**, 399–412.

Sealy, J.R. and Webb, D.A. (1950) *Arbutus* L. Biological flora of the British Isles. *Journal of Ecology*, **38**, 223–236.

Sestini, G. (1992) Implications of climatic change for the Po Delta and Venice Lagoon, in Jeftic, L., Milliman, J.D. and Sestini, G. (eds), *Climatic Change and the Mediterranean* (United Nations Environment Programme), Arnold, London.

Shackleton, J.C. and van Andel, T.H. (1986) Prehistoric shore environment, shellfish availability, and shellfish gathering at Franchthi, Greece. *Geoarchaeology*, **1**, 127–143.

Shmida, A. and Wilson, M.V. (1985) Biological determinants of species diversity. *Journal of Biogeography*, **12**, 1–20.

Simmons, A.H. (1988) Pygmy hippopotamus and early man in Cyprus. *Nature*, **333**, 554–557.

Smith, A.G. and Woodcock, N.H. (1982) Tectonic synthesis of the Alpine-Mediterranean region: a review, in *Alpine Mediterranean Geodynamics*, American Geophysical Union, *Geodynamics Series*, 7, 15–38.

Smith, B. (1997) Water: a critical resource, in King, R., Proudfoot, L. and Smith, B. (eds), *The Mediterranean: Environment and Society*, Arnold, London.

Smith, B.D. (1995) *The Emergence of Agriculture*, Scientific American Library, New York.

Specht, R. (1979) *Ecosystems of the World*, vol. 9 *Heathland and Related Shrublands*, Elsevier, Amsterdam.

Spiegel-Roy, P. and Goldschmidt, E.E. (1996) *Biology of Citrus*, Cambridge University Press, Cambridge.

Stace, C.A. (1989) Dispersal versus vicariance – no contest! *Journal of Biogeography*, **16**, 201–202.

Stace, C.A. (1991) *New Flora of the British Isles*, Cambridge University Press, Cambridge.

Stanley, D.J. and Warne, A.G. (1993a) Sea level and initiation of Predynastic culture in the Nile delta. *Nature*, **363**, 435–438.

Stanley, D.J. and Warne, A.G. (1993b) Nile delta: recent geological evolution and human impact. *Science*, **260**, 628–634.

Stevenson, A.C. and Harrison, R.J. (1992) Ancient forests in Spain: a model for land-use and dry forest management in south-west Spain from 4000 BC to 1900 AD. *Proceedings of the Prehistoric Society*, **58**, 227–247.

Suc, J.-P. (1984) Origin and evolution of the Mediterranean vegetation and climate in Europe. *Nature*, **307**, 429–432.

Taha, M.F., Harb, S.A., Nagib, M.K. and Tantawy, A.H. (1981) Climate of the Near East, in Takahashi, K. and Arakawa, H. (eds), *World Survey of Climatology*, vol. 9 *Climates of Southern and Western Asia*, Elsevier, Amsterdam.

Taylor, H.C. (1978) Capensis, in Werger, M.J.A. (ed.), *Biogeography and Ecology of Southern Africa*, Junk, The Hague.

Terradas, J. (1999) Holm oak and holm oak forests: an introduction, in Rodà, F., Retana, J., Gracia, C.A. and Bellot, J. (eds), *Ecology of Mediterranean Evergreen Oak Forests*, Springer-Verlag, Berlin.

Theophrastus (1976) *de Causis Plantarum*, translated by Einarson, B. and Link, G.K.K., Heinemann, London.

Theophrastus (1916) *Historia Plantarum*, translated by Hort, A., Heinemann, London.

Thirgood, J.V. (1981) *Man and the Mediterranean Forest: A History of Resource Depletion*, Academic Press, London.

Thomas, A.D., Walsh, R.P.D. and Shakesby, R.A. (1999) Nutrient losses in eroded sediment after fire in eucalyptus and pine forests in the wet Mediterranean environment of northern Portugal. *Catena*, **36**, 283–302.

Thompson, D.J. (1983) Nile grain transport under the Ptolomies, in Garnsey, P., Hopkins, K. and Whittaker, C.R. (eds), *Trade in the Ancient Economy*, Chatto and Windus, London.

Thompson, J.D. (1999) Population differentiation in Mediterranean plants: insights into colonization history and the evolution and conservation of endemic species. *Heredity*, **82**, 229–236.

Thompson, R. and Oldfield, F. (1986) *Environmental Magnetism*, Allen & Unwin, London.

Thornes, J.B. (1998a) Mediterranean desertification, in Mairota, P., Thornes, J.B. and Geeson, N. (eds), *Atlas of Mediterranean Environments in Europe: The Desertification Context*, Wiley, Chichester.

Thornes, J.B. (1998b) Results and prospects, in Mairota, P., Thornes, J.B. and Geeson, N. (eds), *Atlas of Mediterranean Environments in Europe: The Desertification Context*, Wiley, Chichester.

Tomaselli, R. (1981a) Main physiognomic types and geographic distribution of shrub systems related to Mediterranean climates, in di Castri, F., Goodall, D.W. and Specht, R. (eds), *Ecosystems of the World*, vol. 11 *Mediterranean-type Shrublands*, Elsevier, Amsterdam.

Tomaselli, R. (1981b) Relations with other ecosystems: temperate evergreen forests, mediterranean coniferous forests, savannahs, steppes and desert shrublands, in di

Castri, F., Goodall, D.W. and Specht, R. (eds), *Ecosystems of the World*, vol. 11 *Mediterranean-type Shrublands*, Elsevier, Amsterdam.

Trabaud, L. (1987) Natural and prescribed fire: survival strategies of plants and equilibrium in mediterranean ecosystems, in Tenhunen, J.D., Catarino, F.M. and Lange, O.L. (eds), *Plant Response to Stress*, Springer-Verlag, Berlin.

Trabaud, L. (1994) Postfire plant community dynamics in the Mediterranean Basin, in Moreno, J.M. and Oechel, W.C. (eds), *The Role of Fire in Mediterranean-type Ecosystems*, Springer-Verlag, New York.

Trudgill, S.T. (2000) *Terrestrial Biosphere*, Pearson Education, Harlow.

Tucker, G.M. and Evans, M.I. (1997) *Habitats for Birds in Europe*, Birdlife International, Cambridge.

Turrill, W.B. (1929) *The Plant-Life of the Balkan Peninsula*, Oxford University Press, Oxford.

Tutin, T.G., Burges, N.A. and Chater, A.O. (1992) *Flora Europaea*, vol. 1, Cambridge University Press, Cambridge.

Tutin, T.G. *et al.* (1964–1980) *Flora Europaea*, 5 volumes, Cambridge University Press, Cambridge.

Tzedakis, P.C. (1993) Long-term tree populations in northwest Greece through multiple Quaternary climatic cycles. *Nature*, **364**, 437–440.

Tzedakis, P.C. (1999) The last climatic cycle at Kopais, central Greece. *Journal of the Geological Society*, **156**, 425–433.

Tzedakis, P.C., Andrieu, V., de Beaulieu, J.-L. and 8 others (1997) Comparison of terrestrial and marine records of changing climate of the last 500,000 years. *Earth and Planetary Science Letters*, **150**, 171–176.

Udias, A. (1985) Seismicity of the Mediterranean Basin, in Stanley, D. and Wezel, F.C. (eds), *Geological Evolution of the Mediterranean Basin*, Springer-Verlag, New York.

UNESCO-FAO (1963) *Bioclimatic Map of the Mediterranean Zone*, UNESCO, Paris.

UNESCO-FAO (1970) *Vegetation Map of the Mediterranean Zone*, UNESCO, Paris.

van Andel, T.H. (1989) Late Quaternary sea-level changes and archaeology. *Antiquity*, **63**, 733–745.

van Andel, T.H. (1994) *New Views on an Old Planet*, 2nd edition, Cambridge University Press, Cambridge.

van Andel, T.H. (1998) Paleosols, red sediments, and the Old Stone Age in Greece. *Geoarchaeology: An International Journal*, **13**, 361–390.

van Andel, T.H. and Runnels, C.N. (1995) The earliest farmers in Europe. *Antiquity*, **69**, 481–500.

van Andel, T.H., Zangger, E. and Demitrack, A. (1990a) Land use and soil erosion in prehistoric and historical Greece. *Journal of Field Archaeology*, **17**, 379–396.

van Andel, T.H., Zangger, E. and Perissoratis, C. (1990b) Quaternary transgressive/regressive cycles in the Gulf of Argos, Greece. *Quaternary Research*, **34**, 317–329.

Vega, J.A. and Diaz-Fierros, F. (1987) Wild fire effects on soil erosion. *Ecologia Mediterranean*, **13**, 119–125.

Vigne, J.D. (1992) Zooarchaeology and biogeographical history of the mammals of Corsica and Sardinia since the last ice age. *Mammal Review*, **22**, 87–96.

Vita-Finzi, C. (1969) *The Mediterranean Valleys*, Cambridge University Press, Cambridge.

Vita-Finzi, C. (1976) Diachronism in Old World alluvial sequences. *Nature*, **263**, 218–219.

Vita-Finzi, C. (1986) *Recent Earth Movements*, Academic Press, London.

Vita-Finzi, C. and King, G.C.P. (1985) The seismicity, geomorphology and structural evolution of the Corinth areas of Greece. *Philosophical Transactions of the Royal Society of London Series A*, **314**, 379–407.

Wagstaff, M. (1981) Buried assumptions: some problems in the interpretation of the 'Younger Fill' raised by data from Greece. *Journal of Archaeological Science*, **8**, 247–264.

Walker, B.H. (1992) Biodiversity and ecological redundancy. *Conservation Biology*, **6**, 18–23.

Walter, K.S. and Gillett, H.J. (1998) *1997 IUCN Red List of Threatened Plants*, IUCN Species Survival Commission, IUCN–The World Conservation Union, Gland, Switzerland and Cambridge, UK.

Wheeler, D. (1991) Majorca's severe storms of September 1989: a reminder of Mediterranean uncertainty. *Weather*, **46**, 21–26.

Whittaker, R.J. (1998) *Island Biogeography*, Oxford University Press, Oxford.

Wigley, T.M.L. (1992) Future climate of the Mediterranean Basin with particular emphasis on changes in precipitation, in Jeftic, L., Milliman, J.D. and Sestini, G. (eds), *Climatic Change and the Mediterranean* (United Nations Environment Programme), Arnold, London.

Wigley, T.M.L. and Farmer, G. (1982) Climate of the eastern Mediterranean and Near East, in Bintliff, J.L. and van Zeist, W. (eds), *Palaeoclimates, Palaeoenvironments and Human Communities in the Eastern Mediterranean Region in Later Prehistory*, BAR International Series no. 133, BAR, Oxford.

Willis, K.J. (1994) The vegetational history of the Balkans. *Quaternary Science Reviews*, **13**, 769–788.

Willis, K.J. and Bennett, K.D. (1995) The Neolithic transition – fact or fiction? Palaeoecological evidence from the Balkans. *The Holocene*, **4**, 326–330.

Wise, S.S., Thornes, J.B. and Gilman, A. (1982) How old are the badlands? A case study from south-east Spain, in Bryan, R. and Yair, A. (eds), *Badland Geomorphology*, GeoBooks, Norwich.

Woodward, J.C. (1995) Patterns of erosion and suspended sediment yield in Mediterranean river basins, in Foster, I.D.L., Gurnell, A.M. and Webb, B.W. (eds), *Sediment and Water Quality in River Catchments*, Wiley, Chichester.

Woodward, J.C., Lewin, J. and Macklin, M.G. (1995) Glaciation, river behaviour and Palaeolithic settlement in upland northwest Greece, in Lewin, J., Macklin, M.G. and Woodward, J.C. (eds), *Mediterranean Quaternary River Environments*, Balkema, Rotterdam.

Wrathall, J.E. (1985) The mistral and forest fires in Provence, Southern France. *Weather*, **40**, 119–124.

Yaalon, D.H. (1997) Soils in the Mediterranean region: what makes them different? *Catena*, **28**, 157–169.

Yassoglou, N. (1998) Mediterranean soils in a desertification context, in Mairota, P., Thornes, J.B. and Geeson, N. (eds), *Atlas of Mediterranean Environments in Europe: The Desertification Context*, Wiley, Chichester.

Zohary, D. and Hopf, M. (1993) *Domestication of Plants in the Old World*, Clarendon Press, Oxford.

Zunino, F. (1981) Dilemma of the Abruzzo bears. *Oryx*, **16**, 153–156.

Index

Note: Individual entries are organised in the sequence: figures (in *bold italics*), tables (in **bold**), boxed material (**bold Box**) and Plates (Plate)

253